T0091840

Solar Flare Magnetic Fields and Plasmas

Yuhong Fan · George Fisher

Editors

Solar Flare Magnetic Fields and Plasmas

Previously published in *Solar Physics* Volume 277, Issue 1, 2012

Editors
Yuhong Fan
High Altitude Observatory
National Center for Atmospheric Research
Boulder, CO, USA

George Fisher
Space Sciences Lab
University of California
Berkeley, CA, USA

ISBN 978-1-4614-3760-4
Springer New York Heidelberg Dordrecht London

Library of Congress Control Number: 2012934840

©Springer Science+Business Media New York 2012
This work is subject to copyright. All rights are reserved by the Publisher, whether the whole or part of the material is concerned, specifically the rights of translation, reprinting, reuse of illustrations, recitation, broadcasting, reproduction on microfilms or in any other physical way, and transmission or information storage and retrieval, electronic adaptation, computer software, or by similar or dissimilar methodology now known or hereafter developed. Exempted from this legal reservation are brief excerpts in connection with reviews or scholarly analysis or material supplied specifically for the purpose of being entered and executed on a computer system, for exclusive use by the purchaser of the work. Duplication of this publication or parts thereof is permitted only under the provisions of the Copyright Law of the Publisher's location, in its current version, and permission for use must always be obtained from Springer. Permissions for use may be obtained through RightsLink at the Copyright Clearance Center. Violations are liable to prosecution under the respective Copyright Law.
The use of general descriptive names, registered names, trademarks, service marks, etc. in this publication does not imply, even in the absence of a specific statement, that such names are exempt from the relevant protective laws and regulations and therefore free for general use.
While the advice and information in this book are believed to be true and accurate at the date of publication, neither the authors nor the editors nor the publisher can accept any legal responsibility for any errors or omissions that may be made. The publisher makes no warranty, express or implied, with respect to the material contained herein.

Cover illustration: Cover photo by NASA/SDO, courtesy of the AIA science team

Printed on acid-free paper

Springer is part of Springer Science+Business Media (www.springer.com)

"The Origin, Evolution and Diagnosis of Solar Flare Magnetic Fields and Plasmas: Honoring the Contributions of Dick Canfield" conference was held at the Center Green Campus of the National Center for Atmospheric Research in Boulder, Colorado, USA from 9 – 11 August 2010. The participants are as follows:

1: Chris Lowder; 2: Ben Lynch; 3: Charlie Lindsey; 4: Jack Campbell; 5: Tom Schad; 6: Ron Moore; 7: Spiro Antiochos; 8: Martin Woodard; 9: Rudy Komm; 10: KD Leka; 11: Alexei Pevtsov; 12: Adam Kowalski; 13: Dana Longcope; 14: Bill Abbett; 15: Alexander Pevtsov; 16: Maria Kazachenko; 17: Mitch Berger; 18: Jiong Qiu; 19: Hugh Hudson; 20: Lindsay Fletcher; 21: Dick Canfield; 22: George Fisher; 23: Maria Weber; 24: Mei Zhang; 25: Anna Malanushenko; 26: Elena Pevtsov; 27: Yuhong Fan.

Contents

Solar Phys (2012) 277:1–2
DOI 10.1007/s11207-011-9913-4

Preface

Yuhong Fan · George Fisher · John Leibacher

Received: 9 November 2011 / Accepted: 9 November 2011 / Published online: 9 December 2011
© The Author(s) 2011. This article is published with open access at Springerlink.com

This Topical Issue of *Solar Physics*, devoted to the dynamics and diagnostics of solar magnetic fields and plasmas, was inspired by a workshop honoring Richard C. (Dick) Canfield. Dick has been making profound contributions to these areas of research over a long and productive scientific career. Many of the articles in this topical issue were first presented as talks during this workshop and represent substantial original work. The workshop was held 9–11 August 2010, at the Center Green campus of the National Center for Atmospheric Research (NCAR) in Boulder, Colorado, with a reception held at the beautiful NCAR Mesa Lab.

Dick Canfield touched the lives and careers of many of today's active members of the solar-physics community, through his role as colleague, teacher, mentor, and as advisor of undergraduates, graduate students, and postdoctoral fellows. He is an enthusiastic participant in, and advocate for, the study of solar physics, and has particularly keen interests in the topics

Richard C. (Dick) Canfield

Solar Flare Magnetic Fields and Plasmas
Guest Editors: Y. Fan and G.H. Fisher

Y. Fan (✉)
High Altitude Observatory, National Center for Atmospheric Research, Boulder, CO, USA
e-mail: yfan@ucar.edu

G. Fisher
Space Sciences Laboratory, University of California, Berkeley, CA, USA

J. Leibacher
National Solar Observatory, Tucson, AZ, USA

J. Leibacher
Institut d'Astrophysique Spatiale, Orsay, France

that are represented in this topical issue. On behalf of the participants of the workshop and the authors of the research articles in this topical issue of *Solar Physics*, we dedicate this topical issue to Dick Canfield for his contributions to science and to the solar-physics community.

The organizers gratefully acknowledge support for this meeting from the US National Science Foundation (NSF) through a supplemental request to grant ATM0551084 at the University of California-Berkeley. We thank Paul Bellaire, at the NSF, for his assistance, as well as Stan Solomon and Michael Thompson for hosting the meeting at NCAR/HAO.

Open Access This article is distributed under the terms of the Creative Commons Attribution Noncommercial License which permits any noncommercial use, distribution, and reproduction in any medium, provided the original author(s) and source are credited.

Solar Phys (2012) 277:3–20
DOI 10.1007/s11207-011-9817-3

Radiative Cooling in MHD Models of the Quiet Sun Convection Zone and Corona

W.P. Abbett · G.H. Fisher

Received: 3 February 2011 / Accepted: 13 June 2011 / Published online: 26 July 2011
© Springer Science+Business Media B.V. 2011

Abstract We present a series of numerical simulations of the quiet-Sun plasma threaded by magnetic fields that extend from the upper convection zone into the low corona. We discuss an efficient, simplified approximation to the physics of optically thick radiative transport through the surface layers, and investigate the effects of convective turbulence on the magnetic structure of the Sun's atmosphere in an initially unipolar (open field) region. We find that the net Poynting flux below the surface is on average directed toward the interior, while in the photosphere and chromosphere the net flow of electromagnetic energy is outward into the solar corona. Overturning convective motions between these layers driven by rapid radiative cooling appears to be the source of energy for the oppositely directed fluxes of electromagnetic energy.

Keywords Convection · Corona · Magnetic fields · Photosphere · Radiative transfer

1. Introduction

To understand the physics of solar activity, we must understand the magnetic and energetic connection between the Sun's convective envelope and corona. The magnetic fields that mediate or energize most, if not all, solar activity are generated below the visible surface within the turbulent convection zone. Yet most of what we can directly measure originates from the solar atmosphere, where physical conditions are fundamentally different from those of the interior. While helioseismic inversions provide an invaluable window into the physics of the Sun's interior, understanding the physical connection between subsurface features and those observed in the solar atmosphere requires a realistic forward model.

Solar Flare Magnetic Fields and Plasmas
Guest Editors: Y. Fan and G.H. Fisher

W.P. Abbett (✉) · G.H. Fisher
Space Sciences Laboratory, University of California, Berkeley, CA, USA
e-mail: abbett@ssl.berkeley.edu

G.H. Fisher
e-mail: fisher@ssl.berkeley.edu

But what level of realism in a numerical model is necessary to describe the complex magnetic connectivity and energetics of the solar atmosphere lying between the visible surface and the corona? It is of great benefit, for example, to formulate a simple, well-defined problem, and set up an idealized numerical experiment that sheds light on the relevant physical processes in an otherwise complex system. In this way, important progress has been made in our understanding of the physics of magnetic-flux emergence in highly stratified model atmospheres (*e.g.*, Manchester *et al.*, 2004; Murray *et al.*, 2006; Magara, 2006; Galsgaard *et al.*, 2007; Fan, 2009; Archontis and Hood, 2010).

Yet the observed evolution of the photospheric magnetic field is often far more complex, particularly in and around CME- and flare-producing active regions. It is difficult to set up a simple magnetic and energetic configuration and an associated physics-based photospheric boundary condition that can initialize a simulation of the solar atmosphere and faithfully mimic the coronal evolution of a complex active region. If we wish to perform first-principles quantitative studies of phenomena such as eruptive events, the energization of the solar wind, active region decay, the transport of magnetic free energy and helicity into the solar atmosphere, and the physics of coronal heating, it is essential to evolve a turbulent model convection zone and corona within a single, large-scale computational domain.

To achieve this, we must accommodate the fundamental energetics of the system while still retaining the ability to study the interplay between large and small-scale magnetic structures that evolve over different timescales. Clearly, radiative transport plays a critical role in the energy balance of the atmospheric layers that bridge the gap between the visible surface and corona. Surface cooling drives convection, and convective turbulence both generates magnetic field and mediates the flux of magnetic energy that enters the solar atmosphere. Yet the physics of radiative transport can be computationally expensive to treat realistically, even in the context of small-scale domains that do not include the convection zone and corona within a single computational volume. For example, energetically important transitions in the solar chromosphere are often decoupled from the local thermodynamic state of the plasma (a state of non-local thermodynamic equilibrium, or non-LTE) suggesting that a truly realistic numerical model must also couple the macroscopic radiative transfer and level population equations to the system of conservation equations (*e.g.*, McClymont and Canfield, 1983; Fisher, Canfield, and McClymont, 1985; Carlsson and Stein, 1992; Abbett and Hawley, 1999; Allred *et al.*, 2005). To complicate matters further, non-thermal physics, and the physics of ion–neutral drag may substantially affect the energy balance of the chromosphere (Krasnoselskikh *et al.*, 2010).

While it remains impractical to perform large-scale, 3D, non-LTE radiative MHD calculations without employing substantial approximations to make the system tractable, it is now common practice to realistically treat optically thick surface cooling in the upper convection zone and photosphere in LTE (a good approximation in these layers). There are many examples of thin-layer, high-resolution calculations that incorporate solutions to the nongray radiative transfer equation in Cartesian domains that include the upper convection zone and extend into the low chromosphere (Bercik, 2002; Stein and Nordlund, 2006; Georgobiani *et al.*, 2007; Rempel, Schüssler, and Knölker, 2009; Cheung *et al.*, 2010). In addition, calculations that realistically treat radiative transfer have been applied to simulations of solar granulation in relatively small-scale domains that also include a transition region and corona (Martínez-Sykora, Hansteen, and Carlsson, 2008; Martínez-Sykora, Hansteen, and Carlsson, 2009; Carlsson, Hansteen, and Gudiksen, 2010).

Our goal, however, is to expand the size of such computational domains to active-region or even global spatial scales while still retaining as realistic a thermodynamic environment as is feasible. Thus, we strive to develop the simplest model possible that allows us to capture the essential physics of the convection-zone-to-corona system while still maintaining the computational efficiency of models in which optically thick radiative cooling is treated in a parameterized fashion (*e.g.*, Abbett, 2007; Fang *et al.*, 2010a). In this way, we hope to make practical the performance of physics-based, first principles simulations, allowing for quantitative, parameter-space studies of processes such as filament formation, active region emergence and decay, and flare and CME initiation.

To simultaneously evolve a realistic model convection zone and corona at any spatial scale presents a number of daunting challenges. The upper convection zone and low solar atmosphere are highly stratified – average thermodynamic quantities change by many orders of magnitude as the domain transitions from a relatively cool, turbulent regime below the visible surface, to a hot, magnetically dominated and shock-dominated regime high in the corona. The physics of the gas transitions from a high-β plasma where the magnetic field is advected by the gas (away from strong active-region complexes) to a low-β regime where the gas is constrained to move along magnetic-field lines. In addition, the radiation field transitions from being optically thick to optically thin. Temporal and spatial scales are highly disparate. Large concentrations of magnetic flux are compressed within intergranular lanes and evolve at convective-turnover timescales, while large coronal loops form and persist for days as active regions emerge and evolve over a course of many months. In addition, the large-scale magnetic structure of the corona can change in a fraction of a second, as small-scale localized magnetic reconnection suddenly reorganizes the large-scale field, often triggering eruptive events along the way.

The corona presents particular challenges. It is well known that in order to accurately reflect the thermodynamics of this region, a model should include the effects of electron heat conduction along magnetic-field lines and radiative cooling in the optically thin "coronal approximation". In addition, some physics-based (*e.g.*, Joule heating) or empirically based source of coronal heating must be present (often introduced at the lower photospheric boundary) if the model corona is to remain hot. But to generate a realistic magnetic carpet, and to study the interaction of granular convection with coronal structures, requires there to be a turbulent model convection zone, and therefore some form of optically thick surface cooling.

In Abbett (2007) we introduced this physics into a 3D MHD convection-zone-to-corona model in the simplest, most computationally efficient way possible – we simply ignored the optically thick radiative-transfer equation entirely, and instead used a parameterized Newton cooling function carefully calibrated against smaller-scale, more realistic radiative-MHD models of magneto-convection where the frequency-dependent LTE transfer equation was solved along with the MHD system (Bercik, 2002).

This approach has been successful in studying the structure of quiet-Sun magnetic fields and active-region flux emergence (Abbett, 2007; Fang *et al.*; 2010a, 2010b). Yet this treatment, while computationally efficient, has a number of limitations. Its principle drawback is that it is ultimately *ad hoc* and requires other, more realistic, simulations as a basis for calibration in order to get meaningful results. The simplified cooling is imposed at a particular height or over a range of gas density, and is not generated in a physical way as a function of optical depth. To address these limitations, we build upon a technique introduced by Abbett and Fisher (2010), and in Section 3 derive a simple, flux-conservative approximation to optically thick cooling that is based on the gray radiative-transfer equation in LTE. We then incorporate a form of this efficient, physics-based approximation into the RADMHD convection-zone-to-corona model of Abbett (2007), which we briefly describe in Section 2. In Section 4 we present new models of an open-field coronal hole region, and study the transport of magnetic energy from below the surface into the corona. Finally, in Section 5 we summarize our results.

2. Numerical Methodology

The parallel code RADMHD solves the following MHD conservation equations semi-implicitly on a three-dimensional Cartesian mesh:

$$\frac{\partial \rho}{\partial t} + \nabla \cdot (\rho \mathbf{u}) = 0, \tag{1}$$

$$\frac{\partial \rho \mathbf{u}}{\partial t} + \nabla \cdot \left[\rho \mathbf{u}\mathbf{u} + \left(p + \frac{B^2}{8\pi} \right) \mathbf{I} - \frac{\mathbf{B}\mathbf{B}}{4\pi} - \mathbf{\Pi} \right] = \rho \mathbf{g}, \tag{2}$$

$$\frac{\partial \mathbf{B}}{\partial t} + \nabla \cdot (\mathbf{u}\mathbf{B} - \mathbf{B}\mathbf{u}) = -\nabla \times (\eta \nabla \times \mathbf{B}), \tag{3}$$

$$\frac{\partial e}{\partial t} + \nabla \cdot (e\mathbf{u}) = -p \nabla \cdot \mathbf{u} + \frac{\eta}{4\pi} |\nabla \times \mathbf{B}|^2 + \Phi + Q. \tag{4}$$

The components of the state vector have the usual definitions: ρ, $\mathbf{u}$, e, p, $\mathbf{B}$, and $\mathbf{g}$ denote the gas density, velocity, internal energy per unit volume, gas pressure, magnetic field, and gravitational acceleration respectively. Here, we assume Gaussian units. The viscous stress tensor is assumed to be of the form $\Pi_{ij} = 2\rho\nu[D_{ij} - 1/3(\nabla \cdot \mathbf{u})\delta_{ij}]$, where $D_{ij} = 1/2(\partial u_i/\partial x_j + \partial u_j/\partial x_i)$ and δ_{ij} denotes the Kronecker δ function. The function $\Phi = \sum_{i,j} \Pi_{ij} D_{ij}$ represents the rate of energy dissipation through viscous diffusion, and ν and η refer to the coefficients of kinematic viscosity and magnetic diffusivity, respectively. These coefficients are assumed constant, and are set to values that correspond to the grid-scale viscous and resistive dissipation. The source term Q includes important energy sources and sinks such as radiative cooling, the divergence of the electron heat flux (in the portion of the domain representing the model transition region and corona), and any desired empirically based coronal-heating function. A complete discussion of the components of this energy source term is provided by Abbett (2007). The system is closed with a non-ideal equation of state, using tabular data provided by the OPAL project (Rogers, 2000). In this article, the portion of the domain corresponding to the corona is heated by the empirically based coronal-heating function described in Abbett (2007), and the effects of Joule dissipation within this region are ignored.

The semi-implicit numerical scheme is parallelized on a domain-decomposed mesh, and the core technique is based on operator splitting with a high-order Crank–Nicholson temporal discretization. We treat the electron thermal conduction, viscous and Joule dissipation, and radiative losses implicitly using a Jacobian-Free Newton–Krylov (JFNK) solver, and require that the remainder of the system be treated explicitly using the Central Weighted Essentially Non-Oscillatory (CWENO) method of Kurganov and Levy (2000) and Balbás and Tadmor (2006). In this way, we remain Courant limited by the magnetosonic wavespeed, and can follow the dynamics of the system in a reliable way (we may choose to relax this constraint when evolving active region magnetic fields over longer timescales). Any local divergence error introduced into the magnetic field as a result of the CWENO central scheme is dissipated by adding an additional artificial source term proportional to $\nabla(\nabla \cdot \mathbf{B})$ to the induction equation. A detailed description of the numerical methodology employed by RADMHD can be found in Section 2 of Abbett (2007).

A number of enhancements and improvements have been incorporated into the RADMHD source code since its initial release in 2007. Most improvements are in the form of improved performance and robustness, better MPI load balancing and scaling, and other enhancements in the code's speed and efficiency. Among the enhancements are:

i) A simplified and improved table inversion and interpolation algorithm that is necessary to incorporate the OPAL data into the code's non-ideal equation of state and the CHIANTI data (Young *et al.*, 2003) into the code's treatment of optically thin radiative cooling.

ii) A new adaptive error algorithm in the GMRES (Generalized Minimum RESidual) sub-step of the JFNK solver that greatly improves convergence rates.

iii) A more robust, global, non-linear CWENO weighting scheme in the explicit substep of RADMHD.

iv) An option to evolve $\log \rho$ rather that ρ itself via the following rewrite of Equation (1):

$$\frac{\partial \ln \rho}{\partial t} + \nabla \cdot (\mathbf{u} \ln \rho) = (\ln \rho - 1) \nabla \cdot \mathbf{u}. \tag{5}$$

Since the model atmosphere is highly stratified, this is often useful as a means of making the code more robust, while at the same time retaining the desired shock-capture characteristics of the numerical scheme. More details on these and other algorithmic improvements will be provided in a technical document under preparation for inclusion with the next release of the code.

In essence, however, the core numerical methods of RADMHD remain the same as those presented in Abbett (2007). In this article, we focus on the portion of the energy source term Q of Equation (4) that contains the approximation for optically thick radiative cooling.

3. An Approximate Treatment of Optically Thick Cooling

Radiative cooling drives surface convection and is a crucial contributor to the energy balance in the region of the solar atmosphere bridging the convection zone and corona. Yet a full frequency-dependent solution to the LTE radiative transfer equation can be computationally expensive for large-scale convection-zone-to-corona calculations, particularly for active region or filament models where timescales are such that the radiative cooling must be updated at intervals close to the MHD CFL limit. Here, we build upon the approach introduced by Abbett and Fisher (2010), and derive an approximate, frequency-integrated expression for optically thick radiative cooling that is based on the gray transfer equation in LTE. We begin by considering the net cooling rate for a volume of plasma at a particular location in the solar atmosphere:

$$R = \int d\Omega \int d\nu \, (\eta_\nu - \kappa_\nu I_\nu). \tag{6}$$

Here, Ω represents solid angle, and ν the frequency. The subscript ν indicates that the emissivity, opacity, and specific intensity [η_ν, κ_ν, and I_ν respectively] depend on frequency. If we define the source function [S_ν] as the ratio of the emissivity to opacity, and rearrange the order of integration, we can recast the net cooling rate in the following form:

$$R = \int d\nu \, \kappa_\nu \int d\Omega \, (S_\nu - I_\nu). \tag{7}$$

We define the mean intensity as $J_\nu \equiv (1/4\pi) \int d\Omega \, I_\nu$ and note that the source function is independent of direction. This allows Equation (7) to be expressed as

$$R = 4\pi \int d\nu \, \kappa_\nu \, (S_\nu - J_\nu). \tag{8}$$

If we now assume a locally plane-parallel geometry, the formal solution for the specific intensity can be written as (*e.g.*, Mihalas, 1978)

$$I_\nu(\tau_\nu, \mu) = \int_0^\infty d\tau' \frac{e^{-|\tau_\nu - \tau'|/|\mu|}}{|\mu|} S_\nu(\tau'), \tag{9}$$

where μ is the cosine of the angle of emergence and τ_ν to the frequency-dependent optical depth. We can now recast the expression for the mean intensity in terms of an integral over optical depth and the cosine of the angle of emergence,

$$J_\nu(\tau_\nu) = \frac{1}{2} \int_0^\infty d\tau' S_\nu(\tau') \int_0^1 d|\mu| \frac{e^{-|\tau_\nu - \tau'|/|\mu|}}{|\mu|}. \tag{10}$$

This allows the integral over μ to be evaluated and expressed in terms of an exponential integral function,

$$J_\nu(\tau_\nu) = \frac{1}{2} \int_0^\infty d\tau' S_\nu(\tau') E_1(|\tau_\nu - \tau'|). \tag{11}$$

Up to now, no approximation other than an assumption of a locally plane-parallel geometry has been made. We now follow the analysis of Abbett and Fisher (2010) and note that the first exponential integral function $[E_1(|\tau_\nu - \tau'|)]$ in Equation (11) is singular when $\tau' = \tau_\nu$, and that this singularity is integrable. Since E_1 is peaked around τ_ν, contributions from $S_\nu(\tau')$ will be centered around $S_\nu(\tau_\nu)$. Thus, to lowest order we can approximate the mean intensity by

$$J_\nu(\tau_\nu) \approx \frac{1}{2} S_\nu(\tau_\nu) \int_0^\infty d\tau' E_1(|\tau_\nu - \tau'|). \tag{12}$$

The integral over optical depth is now easily evaluated, and the result is expressed in terms of the second exponential integral function $[E_2]$:

$$J_\nu(\tau_\nu) \approx S_\nu(\tau_\nu) \left(1 - \frac{E_2(\tau_\nu)}{2}\right). \tag{13}$$

We now rearrange the terms in the above equation, and substitute $1 - J_\nu(\tau_\nu)/S_\nu(\tau_\nu) \approx E_2(\tau_\nu)/2$ into Equation (8), to arrive at an approximation for the net cooling rate:

$$R \approx 2\pi \int d\nu \, \kappa_\nu S_\nu E_2(\tau_\nu). \tag{14}$$

If we further assume LTE, the source function can be expressed as the Planck function $[B_\nu(T)]$ coupling the cooling rate to the local temperature of the plasma $[T]$:

$$R \approx 2\pi \int d\nu \, \kappa_\nu B_\nu(T) E_2(\tau_\nu). \tag{15}$$

We now integrate Equation (15) over frequency. Since $E_2(\tau_\nu)$ is bounded below by zero and above by unity, the integral in Equation (15) obeys this set of inequalities:

$$2\bar{\kappa}\sigma T^4 > 2\pi \int d\nu \, \kappa_\nu B_\nu(T) E_2(\tau_\nu) > 0, \tag{16}$$

where $\bar{\kappa}$ is the Planck-weighted mean opacity.

Because of the range of the $E_2(x)$ function, we can use this inequality to write the integral in Equation (15) in the form $R(\bar{\tau}) = 2\bar{\kappa}C(\bar{\tau})\sigma T^4 E_2(\alpha(\bar{\tau})\bar{\tau})$, where in general, α is a positive, unknown function of mean optical depth $\bar{\tau}$ $[d\bar{\tau} \equiv -\bar{\kappa}\,dz]$, and $C(\bar{\tau})$ is an unknown, $\bar{\tau}$-dependent normalization constant. However, since we expect a close relationship between the mean optical depth $\bar{\tau}$ and the local mean opacity $\bar{\kappa}$, we therefore make the ansatz that α is a constant, but with an unknown value. The expression for R can then be written

$$R \approx 2C\,\bar{\kappa}\sigma T^4 E_2(\alpha\bar{\tau}), \tag{17}$$

where C now represents a $\bar{\tau}$-independent normalization constant of integration.

To determine the normalization constant C, we integrate our cooling function from zero to infinity in optical depth over an isothermal slab to obtain the total radiative flux. The resulting expression must be equal to the known result $F_{\text{tot}} = \sigma T^4$, thus requiring $C = \alpha$. To calibrate α, we compare the cooling rate as a function of depth in test models using this approximation against more realistic models of magneto-convection where the frequency-dependent transfer equation is solved in detail (Bercik, 2002). We conclude that the best-fit value is $\alpha = 1$ (see Figure 1 of Abbett and Fisher, 2010). This implies that optical depth in highly stratified atmospheres is dominated by the local opacity.

With the parameter α specified, we arrive at an approximation for optically thick surface cooling,

$$R \approx 2\bar{\kappa}\sigma T^4 E_2(\bar{\tau}). \tag{18}$$

This expression can be efficiently evaluated at each iteration of an MHD calculation, and we have implemented this volumetric cooling rate as a part of the cell-centered source term Q in Equation (4) (the cooling rate R being a negative heating rate Q).

It is possible for the computational grid to be of sufficient resolution to resolve the local pressure scale heights of a highly stratified model atmosphere while at the same time being poorly resolved in optical depth. This has the potential to lead to numerical error in the calculation of the local cooling rate such that the total radiative flux may not be conserved. We therefore consider a flux-conservative formulation similar to the constrained transport schemes common to many MHD codes (see Stone and Norman, 1992).

We begin by defining a frequency-independent, discretized, optical depth for each iteration where the MHD state variables are updated. Since our simple approximation is based on the assumption of a locally plane-parallel geometry, and we are neglecting (for now) the effects of sideways transport, all that is required is an integration along the vertical direction (*i.e.*, in the direction of the gravitational acceleration). Our discretized expression takes the form

$$\tau_{i,j,k-1/2} \equiv \sum_{n=k_{\text{top}}}^{k} \bar{\kappa}_{i,j,k}(z_{n+1/2} - z_{n-1/2}). \tag{19}$$

Here, the grid coordinates i, j, k are defined at cell centers (consistent with the centralized numerical scheme implemented in RADMHD), and the optical depth is defined at face centers of the mesh cell's control volume perpendicular to the z-direction (we now drop the overbar notation, since the above definition makes it clear that τ is a frequency-averaged quantity). We use tabular, Planck-weighted, mean opacities $[\bar{\kappa}_{i,j,k}]$ provided by the opacity project (Seaton, 2005). The coordinate k_{top} refers to the first ghost cell of the upper coronal boundary of the simulation domain, although in practice it is set to an interior cell bounding the portion of the domain that represents the optically thin corona. Either way, it is presumed that $\tau_{i,j,k_{top}-1/2} = 0$. Note that optical depth increases inward into the atmosphere in the opposite sense of the height z, which is defined to increase outward from the interior toward the visible surface (*i.e.*, $d\tau \equiv -\bar{\kappa}\,dz$).

The radiative cooling of Equation (18) can be expressed in terms of a divergence of a radiative flux. Our treatment of radiative transfer assumes a locally plane-parallel geometry, thus we need only consider the radiative flux at the faces of control volumes normal to the vertical direction. This implies that any horizontal divergence of the radiative flux is assumed negligible when compared to gradients in the vertical direction. The physical justification for this simplification is that changes in emissivity and opacity are generally much greater in the vertical direction of a highly stratified atmosphere than those expected in the transverse direction. This assumption will likely not be valid at the edges of sunspots where the lateral emissivity and opacity gradients are expected to be large. Given this simplification, the divergence of the radiative flux can be expressed as

$$\frac{\partial F}{\partial z} = 2\bar{\kappa}\sigma T^4 E_2(\tau). \tag{20}$$

We now cast this expression in terms of optical depth $d\tau \equiv -\bar{\kappa}\,dz$:

$$\frac{\partial F}{\partial \tau} = -2\sigma T^4 E_2(\tau), \tag{21}$$

and note that this equation is of the form $Q(\tau) = -f(\tau)E_2(\tau)$ with $f(\tau) \equiv 2\sigma T^4$. We now approximate $f(\tau)$ with a Taylor-series expansion centered about τ_k accurate up to second order, and reorder the terms so that the expression is of the form $f(\tau) = A + B\tau$:

$$f(\tau) = \left[f(\tau_k) - \tau_k f'(\tau_k) \right] + \tau f'(\tau_k). \tag{22}$$

Here, $f(\tau_k) = 2\sigma T^4(\tau_k)$ refers to the function $f(\tau)$ evaluated at cell-center coordinate (i, j, k), $f'(\tau_k)$ refers to the function's vertical derivative with respect to optical depth evaluated at the same location, and the constants A and B have the form $A = [f(\tau_k) - \tau_k f'(\tau_k)]$ and $B = f'(\tau_k)$. For brevity, we have dropped the i and j subscripts, but note that these expressions are valid for all grid cells at a particular height.

To obtain the discretized form of Equation (21), we integrate over the control volume of the computational cell,

$$F(\tau_{k+1/2}) - F(\tau_{k-1/2}) = -\int_{\tau_{k-1/2}}^{\tau_{k+1/2}} (A + B\tau)E_2(\tau)\,d\tau. \tag{23}$$

This integral can be evaluated using the relation $dE_n(\tau)/d\tau = -E_{n-1}(\tau)$ to obtain

$$\begin{aligned}
F(\tau_{k+1/2}) - F(\tau_{k-1/2}) = {} & A\left[E_3(\tau_{k+1/2}) - E_3(\tau_{k-1/2}) \right] \\
& + B\left[\tau_{k+1/2}\,E_3(\tau_{k+1/2}) - \tau_{k-1/2}\,E_3(\tau_{k-1/2}) \right. \\
& \left. + E_4(\tau_{k+1/2}) - E_4(\tau_{k-1/2}) \right].
\end{aligned} \tag{24}$$

All that remains is to define a stencil to evaluate $f(\tau_k)$ and $f'(\tau_k)$ in the expressions for A and B. RADMHD employs a central scheme, and it is desirable to derive a stencil consistent with the formalism of the code. Since the temperature $[T]$ is obtained via a table lookup based on cell-centered values of gas density and internal energy per unit volume, we obtain

10

our interpolation stencil by expanding elements of the state vector $q(\tau)$ in a second-order accurate Taylor series about τ_k,

$$q(\tau) = a + b(\tau - \tau_k) + \frac{c}{2}(\tau - \tau_k)^2, \qquad (25)$$

and enforce the definition of cell-averaged quantities along the z-coordinate axis [again, the (i, j) dependence is implicitly assumed, and $\Delta\tau \equiv \tau_{k+1/2} - \tau_{k-1/2}$ is less than zero],

$$q_k = \frac{1}{\Delta\tau} \int_{\tau_{k-1/2}}^{\tau_{k+1/2}} q(\tau)\,d\tau. \qquad (26)$$

We then expand about each of the points τ_{k+1}, τ_k, and τ_{k-1}; substitute the appropriate form of Equation (25) into Equation (26); then perform the integration over each respective control volume. This yields a system of equations whose solution specifies a, b, and c in terms of known cell averages. The compact stencils are equivalent to those of Abbett (2007), and have the form $a = q_k - (q_{k+1} - 2q_k + q_{k-1})/24$, $b = (q_{k+1} - q_{k-1})/(2\Delta\tau)$, and $c/2 = (q_{k+1} - 2q_k + q_{k-1})/(\Delta\tau)^2$. The second term of a arises from the integration of the second-order term in the Taylor expansion of $q(\tau)$, and ensures that the interpolation scheme maintains second-order accuracy, and that the following discretized, cell-centered forms of A and B have desirable stability properties:

$$A = f_k - \frac{1}{24}(f_{k+1} - 2f_k + f_{k-1}) - \tau_k \frac{1}{2\Delta\tau}(f_{k+1} - f_{k-1}), \qquad (27)$$

$$B = \frac{1}{2\Delta\tau}(f_{k+1} - f_{k-1}). \qquad (28)$$

Here $f_k = 2\sigma T_k^4$, and $\tau_k = \tau_{k-1/2} - \kappa_k(\Delta z/2)$. With A and B specified, we arrive at an expression for a flux-conservative approximation to the optically thick radiative source term,

$$\begin{aligned}
Q(\tau_{i,j,k}) = {}& A\overline{\kappa}_{i,j,k}\left[E_3(\tau_{i,j,k-1/2}) - E_3(\tau_{i,j,k+1/2})\right] \\
& + B\overline{\kappa}_{i,j,k}\left[\tau_{i,j,k-1/2}\,E_3(\tau_{i,j,k-1/2}) - \tau_{i,j,k+1/2}\,E_3(\tau_{i,j,k+1/2})\right. \\
& \left. + E_4(\tau_{i,j,k-1/2}) - E_4(\tau_{i,j,k+1/2})\right].
\end{aligned} \qquad (29)$$

By design, this expression will conserve flux to machine roundoff, as can easily be verified by showing that $Q_k + Q_{k-1} = F_{k+1/2} - F_{k-3/2}$ for all points (i, j, k). Thus, we have two ways of implementing our approximation: the flux-conservative approach of Equation (29), and the non-conservative approach obtained by directly evaluating Equation (18) using cell-centered quantities. The flux-conserving method requires additional table lookups each iteration to evaluate the exponential integrals, but is helpful in cases where the optical-depth scale is not particularly well-resolved.

Our approximation for gray LTE cooling is applied only in those regions of the computational domain where such an approximation is needed. Specifically, we apply the approximation over a range of optical depths that extend from $\tau = 10$ to $\tau = 0.1$. At greater optical depths, we use the diffusion approximation (with tabular Rosseland mean opacities provided by Seaton, 2005), and at smaller optical depths we use the optically thin approximation (using CHIANTI data from Dere *et al.*, 1997 and Young *et al.*, 2003 to specify the optically thin cooling curve) as described in Abbett (2007) and Lundquist, Fisher, and McTiernan (2008).

The simulations we present in Section 4 use the non-conservative technique. The conservative approach described above was motivated by the fact that in some cases, where the model atmosphere is poorly resolved in optical depth, the strong cooling prescribed by Equation (18) can be concentrated in a narrow one- or two-zone layer of a model atmosphere. As a practical matter, this required that we enforce a limit on the maximum amount of cooling per unit mass allowable in any given grid cell. This cooling floor is somewhat artificial, and is not necessary in the flux-conservative method, which has the effect of spreading the cooling over adjoining cells in a more physical way. Another option would be to have a separate grid for optical depth, but we decided against this because of the possibility of introducing additional interpolation error that is difficult to characterize. We are currently testing our new flux-conservative scheme, and plan to fully implement it in a new radiation subroutine for RADMHD that also includes the important effects of sideways transport. We hope to report on these efforts in the near future.

4. A Model of an Open Flux Region

We initiate our calculations using the procedure of Abbett (2007). Briefly, we begin by relaxing a 1D-symmetric average stratification, then expand the domain to three dimensions and break the 1D symmetry by introducing a small, random energy perturbation in the superadiabatically stratified portion of the computational domain representing the solar convection zone. Convective turbulence develops as the simulations progress, and we allow the model convection zone to dynamically relax. We show results from two separate simulations: one was performed locally using 112 processors of a relatively small Beowulf cluster, and the other was performed on NASA's Discover supercomputer using 512 processors. Both simulations simultaneously evolve a model convection zone and corona, and each domain has a vertical extent of 12 Mm, with a 2.5 Mm deep model convection zone. The development model that was run on the local Beowulf cluster has a domain that spans $21 \times 12 \times 12$ Mm^3 at a resolution of $448 \times 256 \times 256$, while the larger run on Discover spans $24 \times 24 \times 12$ Mm^3 at a relatively high resolution of $512 \times 512 \times 256$. In each case, only ≈ 0.66 percent of the total computational effort was expended by the approximate treatment of the radiative transfer on average within any given MPI subdomain during a given timestep. On the Intel Xeon E5420 CPUs of our local Beowulf cluster, the computing time per update of this substep is approximately 0.03 core-microseconds per point. The simulations presented here should be considered relatively small-scale in the context of the capability of the algorithms presented – the code scales well on multiple processors, and once our development work is complete we intend to dramatically extend the spatial scale of the models.

The simulations presented here differ from those of Abbett (2007) in a fundamental way. The approximation we now use for optically thick cooling eliminates all of the *ad-hoc* calibrated parameters present in the older models. Specifically, the height and magnitude of the optically thick radiative source term is now calculated in a physically self-consistent way based on an optical-depth scale rather than on an specified density or height range attenuated by envelope functions (*cf.* Section 2.1.1 of Abbett, 2007). Once α of Equation (17) has been calibrated against more realistic models, no further adjustment is required, and each atmosphere relaxes to a state determined by the solution of the system of Equations (1) – (4) subject to imposed boundary conditions.

We apply periodic boundaries in the horizontal directions, and a simple, somewhat-artificial, closed lower boundary. Specifically, the internal energy per unit volume within ghost cells adjacent to the domain's lower boundary is set such that a temperature gradient is maintained that best matches the average stratification at a corresponding height in the Bercik (2002) magneto-convection models. In addition, the ghost cells at the lower boundary are specified such that the vertical components of the velocity and magnetic field and the vertical gradients of the horizontal components of the velocity and magnetic field are zero, and such that the gradient of the gas density is maintained. The upper coronal boundary is initially taken to be anti-symmetric during the relaxation procedure (*i.e.*, ghost zones are set such that the vertical component of the velocity and magnetic field vanishes at the boundary while all other components of the MHD state vector maintain a zero vertical gradient across the boundary interface), then is set to a standard zero-gradient boundary condition once magnetic fields are introduced (*i.e.*, ghost cells are set such that all components of the MHD state vector maintain a zero vertical gradient across the upper boundary interface). For the simulations presented here, once the purely hydrodynamic-model convection zone is relaxed, we introduce a weak 1 G vertically directed magnetic field. This is intended to create an open-flux region, such as one might expect within a coronal hole.

As the simulations progress, the convective turbulence acts to stretch and amplify the field, and the portion of the domain representing the corona begins to heat as a result of the magnetic-field-dependent empirically based coronal-heating source term. This heating function is based on the Pevtsov *et al.* (2003) power-law relationship between X-ray luminosity and total unsigned magnetic flux observed at the surface (see Equation (12) of Abbett, 2007). For this study, we are content to rely on empirical heating rather than Joule dissipation to energize the model corona since our focus is on the transport of magnetic energy into the atmosphere, not the heating of the low atmosphere and corona.

Up to the point that magnetic field was introduced into the simulation domain, the model corona was simply an unphysical, cold, nearly evacuated region. Once the corona heats, we activate the implicit electron thermal-conduction source term. This builds a corona, but reduces our timestep somewhat since the stiffness of the system is increased and convergence rates in the JFNK substep can become an issue. Thus, it tends to be the last step of our relaxation process. After several additional turnover times, we begin our analyses.

The left frame of Figure 1 shows the temperature at the visible surface ($\tau = 1$) of a relaxed convection-zone-to-corona model using the new formalism of Section 3. The granular pattern and convective turnover times compare well to the realistic magneto-convection models of Bercik (2002). This, along with the reverse granulation pattern shown in Figure 2, indicates that our approximation to optically thick radiative transfer is capturing the physics of surface cooling at least well enough to generate and sustain solar-like convective features.

The right frame of Figure 1 shows the complex magnetic structure threading a small portion of the simulation domain (as indicated by the cyan box in the upper-right corner of the left frame). Figure 3 shows a larger subdomain at a later time, and more clearly illustrates the characteristics of the magnetic structure. In the region where convective cells turn over, the average plasma-β remains relatively high. As a result, much of the magnetic field remains entrained in the plasma, turns over, and is recirculated back below the surface. Thus, at any given time, this region is filled with horizontally directed field, and that field tends to be less concentrated than is typical of fields entrained in the vortical downdrafts present at the visible surface and below. In addition, the presence of canopy-like structures (where strong concentrations of field above intergranular lanes and photospheric downdrafts open into the upper atmosphere and spread out like a fan) also contribute to the net amount of horizontally directed magnetic field threading the atmosphere below the corona.

The presence of horizontal fields in the atmosphere has consequences for the transport of magnetic energy into the upper atmosphere. Consider the electromagnetic Poynting flux,

$$\mathbf{S} = \frac{1}{4\pi} c \mathbf{E} \times \mathbf{B}. \tag{30}$$

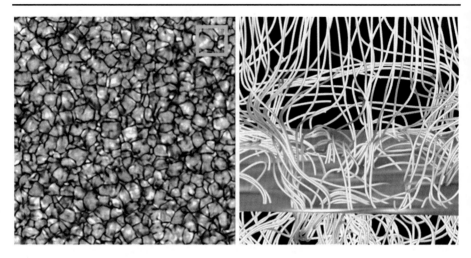

Figure 1 Left: Temperature at the RADMHD model photosphere. Right: Magnetic-field lines threading the low atmosphere over a small subdomain (the box in the left frame indicates the approximate size of the corresponding subdomain). The gray slice indicates the average height of the visible surface. The domain spans $24 \times 24 \times 12$ Mm^3 at a resolution of $512 \times 512 \times 256$.

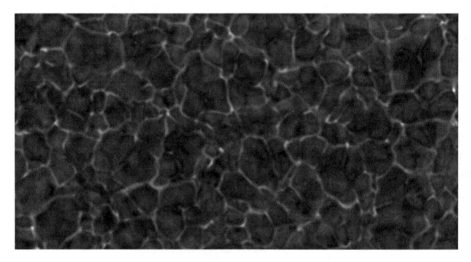

Figure 2 Temperature in the RADMHD low chromosphere showing a reverse granulation pattern. Lighter (darker) colors indicate hotter (cooler) temperatures. In the models, this occurs because the radiative cooling diminishes with height, and the $p\nabla \cdot \mathbf{v}$ work of converging and diverging flows above the intergranular lanes begins to dominate. The horizontal slice spans 21×12 Mm^2 at a resolution of 448×256.

The vertical component of the Poynting flux is a measure of the amount of electromagnetic energy flowing into, or out of the solar atmosphere from below the surface where it is generated. In Figure 4, we display S_z as a grayscale image at two layers in the model atmosphere – dark shades correspond to a flux of magnetic energy directed toward the interior, while lighter shades correspond to an outward-directed flux. The top frame shows the vertical component of Poynting flux along an $x–y$ slice positioned just below the visible surface, and the lower frame shows S_z along a slice positioned in the low atmosphere,

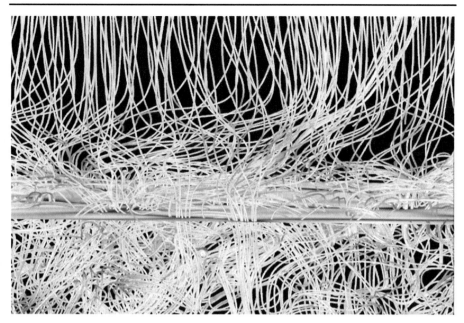

Figure 3 Magnetic-field lines threading the low atmosphere over a portion of the computational domain. The gray slice represents the approximate position of the model's visible surface. Horizontally directed magnetic fields due to the spreading of canopy-like structures and overturning convective cells permeate the low atmosphere.

400 km higher. Careful examination of Figure 4 reveals an imbalance in the outward- and inward-directed flux. Below the surface, there appears to be a net flow of magnetic energy into the convective interior along the strong vortical downdrafts contained within intergranular lanes. Conversely, in the low atmosphere, the vertical component of the Poynting flux appears more diffuse, and there appears to be a net excess of outward-directed flux, particularly within overturning granules.

This is shown more clearly in Figure 5 where S_z is integrated over each layer in the computational domain, and plotted as a normalized quantity as a function of height [z]. The dashed vertical line in the figure represents the average height of the visible surface. What is clear is that magnetic energy on average is directed downward into the interior below the visible surface. It is at the surface and above that the net vertical Poynting flux changes sign and becomes outwardly directed. This suggests that the kinetic motion of overturning granules in the model's overshoot layer provides the source of magnetic energy for the corona, not the deeper layers below the optical surface, where magnetic flux and energy are being pumped down into the interior along intergranular downflows.

This can be understood in a fairly straightforward way. The vertical component of the Poynting flux can be expressed as

$$S_z = \frac{1}{4\pi} (c\mathbf{E}_h \times \mathbf{B}_h) \cdot \hat{\mathbf{z}}, \tag{31}$$

where $\mathbf{E}_h$ and $\mathbf{B}_h$ refer to the horizontal components of the electric and magnetic field respectively. If we assume ideal MHD, then the horizontal component of the electric field can

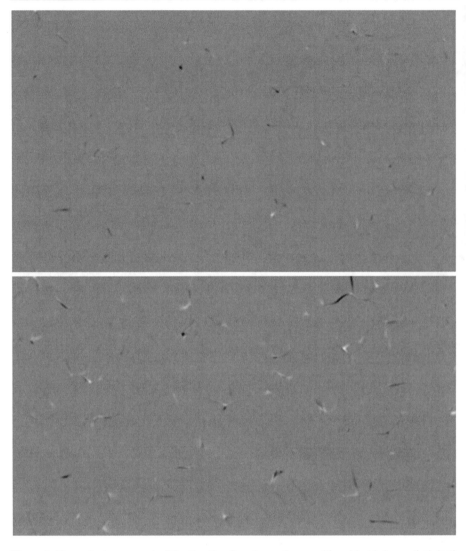

Figure 4 The vertical component of the Poynting flux along a layer positioned just below optical depth unity (top) and 400 km higher near the tops of overturning granules (bottom). Light colors correspond to outward-directed flux (toward the corona); dark colors represent inward-directed flux (toward the convective interior). Each slice spans $21 \times 12 \, \mathrm{Mm}^2$.

be written as follows:

$$c\mathbf{E}_h = -u_z\hat{\mathbf{z}} \times \mathbf{B}_h - \mathbf{u}_h \times B_z\hat{\mathbf{z}}. \tag{32}$$

Just above the visible surface, the magnetic field remains entrained in the fluid as convective cells overturn. At this height, the strongest field concentrations are located near the edges of overturning granules as divergent flows from neighboring cells compress the field. On average, there is more of a contribution to the horizontal electric field from the second term of Equation (32), $c\mathbf{E}_h = -\mathbf{u}_h \times B_z\hat{\mathbf{z}}$, since the magnetic field becomes more vertical as converging flows compress flux into a relatively small area. The contribution of this term to the

Figure 5 The normalized net vertical component of the Poynting flux over a portion of the domain centered at the model's photosphere. The dashed vertical line represents the approximate height of the visible surface. Above the visible surface, electromagnetic energy tends to flow outward toward the corona, while below the surface, energy flows inward toward the convective interior. Above $z \approx 0.5$ (in the model's low chromosphere) the Poynting flux tends to remain outwardly directed, but its magnitude is, on average, less than a percent of its maximum value.

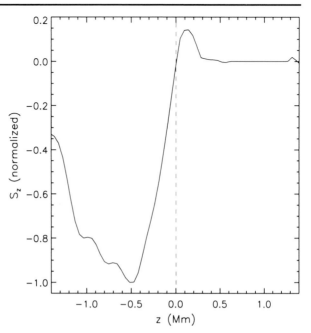

vertical component of the Poynting flux can be expressed as $4\pi S_z = [(-\mathbf{u}_h \times B_z \hat{\mathbf{z}}) \times \mathbf{B}_h] \cdot \hat{\mathbf{z}}$. Simplified, this becomes $4\pi S_z = -B_z(\mathbf{B}_h \cdot \mathbf{u}_h)$.

To illustrate the correlation between the horizontal magnetic fields [$\mathbf{B}_h$] and the converging surface flows [$\mathbf{u}_h$], consider a weak, vertically oriented, untwisted magnetic flux tube that passes through the surface. Suppose it is acted on by a strong converging flow in a thin layer at the surface. If the magnetic field of the tube is oriented in the positive z direction, then just above the surface, the compression will tilt the field lines and create horizontal components of the magnetic field in the opposite direction of the converging flow. If the magnetic field is oriented in the negative z direction, then the horizontal components of the field will be aligned with the flow. Either way, $4\pi S_z = -B_z(\mathbf{B}_h \cdot \mathbf{u}_h)$ is positive above the surface. Obviously, the dynamics of the model are far more complex than this simple thought experiment. Nevertheless, we do find a net positive contribution to the Poynting flux from the second term of Equation (32) along the edges of overturning granules above the surface where the field is being compressed.

Below the photosphere, the situation is quite different. The strongest magnetic fields are concentrated within localized vortical downdrafts. The asymmetry between these strong downdrafts and the broad upwelling plasma in stratified convection is well known, and may provide a mechanism whereby magnetic flux can be pumped into the interior (Tobias *et al.*, 2001). We find more of a contribution to the net Poynting flux from the first term of Equation (32), $4\pi S_z = [(-u_z \hat{\mathbf{z}} \times \mathbf{B}_h) \times \mathbf{B}_h] \cdot \hat{\mathbf{z}}$. This can be expressed more simply as $4\pi S_z = u_z B_h^2$, and in this form it is easy to see that a net downward transport of magnetic flux is consistent with a net downward-directed Poynting flux. This downward-directed flux of electromagnetic energy below the surface is consistent with other simulations of radiative-magneto-convection (*e.g.*, Vögler and Schüssler, 2007).

The first to recognize this change in direction of the flow of electromagnetic energy was Steiner *et al.* (2008) who referred to the visible surface as "a separatrix for the vertically directed Poynting flux". Our results are consistent with their findings, although we conclude that in the larger domain, the upward-directed net flow of magnetic energy tends to arise from the action of the compressive flows of overturning convection as magnetic flux is expelled from cell centers and concentrated into the intergranular regions. However, higher in the model atmosphere as the gas transitions to a low-β regime (the upper chromosphere–transition region boundary), we also see a small buildup of magnetic flux, for reasons similar to those of Steiner *et al.* (2008). Namely, that the dynamic chromosphere transitions to a stable, subadiabatic, magnetically dominated regime, and there is a magnetic reservoir on average as magnetic flux that is advected upward enters the stable regime and does not get recirculated back into the convective interior (similar in some ways to the overshoot layer at the base of the convection zone). This is reflected in the small peak in Figure 5 at a height of ≈ 1.3 Mm above the visible surface. Above this transition region interface in the open field of the model corona, energy is transported via magnetosonic and Alfvén waves. We note that in these simulations, the magnetic field has yet to fully saturate (*i.e.*, there is a small increase in magnetic energy over time as magnetic field is stretched and amplified by convective turbulence). While this indicates that the atmosphere has yet to fully relax, this increase in magnetic energy (and any Joule heating below the corona) is negligible in comparison to the divergence of the Poynting flux and the work done on the magnetic field by convective motions.

In some sense, the height at which the transition between outward and inward flow of electromagnetic energy takes place is less important than the fact that such a transition exists. What the simulations seem to suggest is that, in quiescent regions away from particularly strong concentrations of magnetic flux, there is not a continuous flow of electromagnetic energy from below the surface out into the corona. Instead, the mechanical energy of convection mediates the flow of magnetic energy in the relatively high-β surface layers where the magnetic field remains frozen into the plasma. Of course, in and around very strong concentrations of magnetic flux, the situation is undoubtedly quite different.

5. Discussion and Conclusions

We have developed an approximate treatment of optically thick radiative surface cooling that successfully reproduces the average thermodynamic stratification of smaller-scale, more realistic numerical models where the frequency-dependent radiative transfer equation in LTE is solved in detail. This technique retains the computational efficiency of earlier parameterized methods, but does not require continual calibration against more realistic simulations. We find that with the new method we are able to initiate and sustain a stable convection pattern with a distribution of cell sizes and turnover times characteristic of solar granulation.

The method presents a middle ground between realistic radiative MHD models that solve the transfer equation in detail, and idealized models that simply impose a thermodynamic stratification, or ignore the physics of radiative transport entirely. The motivation for developing this technique is to make feasible physics-based large-scale or global parameter-space studies of the interaction of active region-scale magnetic fields with the small scale fields associated with granular convection in a domain that includes both a convection zone and corona.

Whether the approximate treatment captures enough of the essential physics of the system still remains to be seen. The technique is certainly limited by the fact that sideways transport is ignored, and important physics of the chromosphere has yet to be included in the current models. Even so, we are able to generate solar-like convective turbulence in a physically self-consistent way, and follow the magnetic evolution of structures that thread the interface between the convective interior and corona.

In particular, we presented two simulations that confirm the existence of a "separatrix" in the flow of magnetic energy from the interior to the atmosphere (see Steiner *et al.*, 2008), and demonstrate that it is the mechanical energy of surface convection driven by strong radiative cooling that is the source of energy for this divergent flux of electromagnetic energy. In a quiescent region, in the absence of strong concentrations of magnetic flux, our models suggest that it is the low photosphere that provides the source of electromagnetic energy to the chromosphere and corona, not the subsurface layers.

Acknowledgements This research was funded in part by the NASA Heliophysics Theory Program (grant NNX08AI56G), the NASA Living-With-a-Star TR&T Program (grant NNX08AQ30G), and the NSF's AGS Program (grant ATM-0737836). The authors wish to acknowledge Dick Canfield's pioneering efforts in the development of radiation hydrodynamics during the 1970s and 1980s. The particular description in this article for the simplified radiative cooling treatment was inspired by the very first work (unpublished) that GHF did for Dick Canfield as a graduate student, namely an investigation of escape probability treatments for continuum radiation processes. Indeed, it is possible to derive the radiative-cooling treatment described here using escape probability concepts.

References

Abbett, W.P.: 2007, The magnetic connection between the convection zone and corona in the quiet Sun. *Astrophys. J.* **665**, 1469 – 1488. doi:10.1086/519788.

Abbett, W.P., Fisher, G.H.: 2010, Improving large-scale convection-zone-to-corona models. *Mem. Soc. Astron. Ital.* **81**, 721 – 728.

Abbett, W.P., Hawley, S.L.: 1999, Dynamic models of optical emission in impulsive solar flares. *Astrophys. J.* **521**, 906 – 919. doi:10.1086/307576.

Allred, J.C., Hawley, S.L., Abbett, W.P., Carlsson, M.: 2005, Radiative hydrodynamic models of the optical and ultraviolet emission from solar flares. *Astrophys. J.* **630**, 573 – 586. doi:10.1086/431751.

Archontis, V., Hood, A.W.: 2010, Flux emergence and coronal eruption. *Astron. Astrophys.* **514**, 56 – 59. doi:10.1051/0004-6361/200913502.

Balbás, J., Tadmor, E.: 2006, Nonoscillatory central schemes for one- and two-dimensional magnetohydrodynamics equations. II: High-order semidiscrete schemes. *SIAM J. Sci. Comput.* **28**(2), 533 – 560. doi:10.1137/040610246. http://link.aip.org/link/?SCE/28/533/1.

Bercik, D.J.: 2002, A numerical investigation of the interaction between convection and magnetic field in a solar surface layer. PhD thesis, Michigan State University.

Carlsson, M., Stein, R.F.: 1992, Non-LTE radiating acoustic shocks and Ca II K2V bright points. *Astrophys. J. Lett.* **397**, L59 – L63. doi:10.1086/186544.

Carlsson, M., Hansteen, V.H., Gudiksen, B.V.: 2010, Chromospheric heating and structure as determined from high resolution 3D simulations. *Mem. Soc. Astron. Ital.* **81**, 582 – 587.

Cheung, M.C.M., Rempel, M., Title, A.M., Schüssler, M.: 2010, Simulation of the formation of a solar active region. *Astrophys. J.* **720**, 233 – 244. doi:10.1088/0004-637X/720/1/233.

Dere, K.P., Landi, E., Mason, H.E., Monsignori Fossi, B.C., Young, P.R.: 1997, CHIANTI – an atomic database for emission lines. *Astron. Astrophys. Suppl.* **125**, 149 – 173.

Fan, Y.: 2009, The emergence of a twisted flux tube into the solar atmosphere: sunspot rotations and the formation of a coronal flux rope. *Astrophys. J.* **697**, 1529 – 1542. doi:10.1088/0004-637X/697/2/1529.

Fang, F., Manchester, W., Abbett, W.P., van der Holst, B.: 2010a, Simulation of flux emergence from the convection zone to the corona. *Astrophys. J.* **714**, 1649 – 1657. doi:10.1088/0004-637X/714/2/1649.

Fang, F., Manchester, W.B., Abbett, W.P., van der Holst, B., Schrijver, C.J.: 2010b, Simulation of flux emergence in solar active regions. In: *AGU Fall Meeting Abstracts*, A1781.

Fisher, G.H., Canfield, R.C., McClymont, A.N.: 1985, Flare loop radiative hydrodynamics. V – Response to thick-target heating. VI – Chromospheric evaporation due to heating by nonthermal electrons. VII – Dynamics of the thick-target heated chromosphere. *Astrophys. J.* **289**, 414 – 441. doi:10.1086/162901.

Galsgaard, K., Archontis, V., Moreno-Insertis, F., Hood, A.W.: 2007, The effect of the relative orientation between the coronal field and new emerging flux. I. Global properties. *Astrophys. J.* **666**, 516 – 531. doi:10.1086/519756.

Georgobiani, D., Zhao, J., Kosovichev, A.G., Benson, D., Stein, R.F., Nordlund, Å.: 2007, Local helioseismology and correlation tracking analysis of surface structures in realistic simulations of solar convection. *Astrophys. J.* **657**, 1157 – 1161. doi:10.1086/511148.

Krasnoselskikh, V., Vekstein, G., Hudson, H.S., Bale, S.D., Abbett, W.P.: 2010, Generation of electric currents in the chromosphere via neutral–ion drag. *Astrophys. J.* **724**, 1542 – 1550. doi:10.1088/0004-637X/724/2/1542.

Kurganov, A., Levy, D.: 2000, A third-order semidiscrete central scheme for conservation laws and convection-diffusion equations. *SIAM J. Sci. Comput.* **22**(4), 1461 – 1488.

Lundquist, L.L., Fisher, G.H., McTiernan, J.M.: 2008, Forward modeling of active region coronal emissions. I. Methods and testing. *Astrophys. J. Suppl.* **179**, 509 – 533. doi:10.1086/592775.

Magara, T.: 2006, Dynamic and topological features of photospheric and coronal activities produced by flux emergence in the Sun. *Astrophys. J.* **653**, 1499 – 1509. doi:10.1086/508926.

Manchester, W. IV, Gombosi, T., DeZeeuw, D., Fan, Y.: 2004, Eruption of a buoyantly emerging magnetic flux rope. *Astrophys. J.* **610**, 588 – 596. doi:10.1086/421516.

Martínez-Sykora, J., Hansteen, V., Carlsson, M.: 2008, Twisted flux tube emergence from the convection zone to the corona. *Astrophys. J.* **679**, 871 – 888. doi:10.1086/587028.

Martínez-Sykora, J., Hansteen, V., Carlsson, M.: 2009, Twisted flux tube emergence from the convection zone to the corona. II. Later states. *Astrophys. J.* **702**, 129 – 140. doi:10.1088/0004-637X/702/1/129.

McClymont, A.N., Canfield, R.C.: 1983, Flare loop radiative hydrodynamics. I – Basic methods. *Astrophys. J.* **265**, 483 – 506. doi:10.1086/160692.

Mihalas, D.: 1978, *Stellar Atmospheres*, 2nd edn. San Francisco, Freeman.

Murray, M.J., Hood, A.W., Moreno-Insertis, F., Galsgaard, K., Archontis, V.: 2006, 3D simulations identifying the effects of varying the twist and field strength of an emerging flux tube. *Astron. Astrophys.* **460**, 909 – 923. doi:10.1051/0004-6361:20065950.

Pevtsov, A.A., Fisher, G.H., Acton, L.W., Longcope, D.W., Johns-Krull, C.M., Kankelborg, C.C., Metcalf, T.R.: 2003, The relationship between X-Ray radiance and magnetic flux. *Astrophys. J.* **598**, 1387 – 1391. doi:10.1086/378944.

Rempel, M., Schüssler, M., Knölker, M.: 2009, Radiative magnetohydrodynamic simulation of sunspot structure. *Astrophys. J.* **691**, 640 – 649. doi:10.1088/0004-637X/691/1/640.

Rogers, F.J.: 2000, Ionization equilibrium and equation of state in strongly coupled plasmas. *Phys. Plasmas* **7**, 51 – 58. doi:10.1063/1.873815.

Seaton, M.J.: 2005, Opacity project data on CD for mean opacities and radiative accelerations. *Mon. Not. Roy. Astron. Soc.* **362**, 1 – 3. doi:10.1111/j.1365-2966.2005.00019.x.

Stein, R.F., Nordlund, Å.: 2006, Solar small-scale magnetoconvection. *Astrophys. J.* **642**, 1246 – 1255. doi:10.1086/501445.

Steiner, O., Rezaei, R., Schaffenberger, W., Wedemeyer-Böhm, S.: 2008, The horizontal internetwork magnetic field: numerical simulations in comparison to observations with Hinode. *Astrophys. J. Lett.* **680**, L85 – L88. doi:10.1086/589740.

Stone, J.M., Norman, M.L.: 1992, ZEUS-2D: A radiation magnetohydrodynamics code for astrophysical flows in two space dimensions. II. The magnetohydrodynamic algorithms and tests. *Astrophys. J. Suppl.* **80**, 791 – 818. doi:10.1086/191681.

Tobias, S.M., Brummell, N.H., Clune, T.L., Toomre, J.: 2001, Transport and storage of magnetic Field by overshooting turbulent compressible convection. *Astrophys. J.* **549**, 1183 – 1203. doi:10.1086/319448.

Vögler, A., Schüssler, M.: 2007, A solar surface dynamo. *Astron. Astrophys.* **465**, 43 – 46. doi:10.1051/0004-6361:20077253.

Young, P.R., Del Zanna, G., Landi, E., Dere, K.P., Mason, H.E., Landini, M.: 2003, CHIANTI – an atomic database for emission lines. VI. Proton rates and other improvements. *Astrophys. J. Suppl.* **144**, 135 – 152. doi:10.1086/344365.

Solar Phys (2012) 277:21–29
DOI 10.1007/s11207-011-9839-x

The Multiple Continuum Components in the White-Light Flare of 16 January 2009 on the dM4.5e Star YZ CMi

A.F. Kowalski · S.L. Hawley · J.A. Holtzman ·
J.P. Wisniewski · E.J. Hilton

Received: 2 January 2011 / Accepted: 5 August 2011 / Published online: 4 October 2011
© Springer Science+Business Media B.V. 2011

Abstract The white light during M dwarf flares has long been known to exhibit the broad-band shape of a $T \approx 10\,000$ K blackbody, and the white light in solar-flares is thought to arise primarily from hydrogen recombination. Yet, a current lack of broad-wavelength coverage solar flare spectra in the optical/near-UV region prohibits a direct comparison of the continuum properties to determine if they are indeed so different. New spectroscopic observations of a secondary flare during the decay of a megaflare on the dM4.5e star YZ CMi have revealed multiple components in the white-light continuum of stellar flares, including both a blackbody-like spectrum and a hydrogen-recombination spectrum. One of the most surprising findings is that these two components are anti-correlated in their temporal evolution. We combine initial phenomenological modeling of the continuum components with spectra from radiative hydrodynamic models to show that continuum veiling causes the measured anti-correlation. This modeling allows us to use the components' inferred properties to predict how a similar spatially resolved, multiple-component, white-light continuum might appear using analogies to several solar-flare phenomena. We also compare the properties of the optical stellar flare white light to Ellerman bombs on the Sun.

Keywords White-light flares · Solar-stellar connection · Radiative transfer · Ellerman bombs

1. Introduction

In both solar and stellar flares, the near-UV and optical (white-light) continuum emission is an energetically important but unexplained phenomenon. On the Sun, the white-light con-

Solar Flare Magnetic Fields and Plasmas
Guest Editors: Y. Fan and G.H. Fisher

J.P. Wisniewski is a NSF Astronomy & Astrophysics Postdoctoral Fellow.

A.F. Kowalski (✉) · S.L. Hawley · J.P. Wisniewski · E.J. Hilton
Astronomy Department, University of Washington, Box 351580, Seattle, WA 98195, USA
e-mail: adamfk@u.washington.edu

J.A. Holtzman
Department of Astronomy, New Mexico State University, Box 30001, Las Cruces, NM 88003, USA

tinuum appears in small regions of transient emission that are spatially and temporally co-incident with hard X-ray bursts (Rust and Hegwer, 1975; Hudson *et al.*, 1992; Neidig and Kane, 1993; Fletcher *et al.*, 2007). This relation suggests that the origin of the white light is related to the energy deposited in the lower atmosphere by nonthermal electrons accelerated during flares. Broad-wavelength-coverage spectral observations are sparse and date back to several large solar flares from the 1970s and 1980s (Machado and Rust, 1974; Hiei, 1982; Neidig, 1983; Donati–Falchi, Smaldone, and Falciani, 1984). These spectra are consistent with continua arising primarily from the hydrogen Balmer continuum and H^- emission.

Whereas the largest solar flares emit $< 10^{32}$ ergs in the white-light continuum and last not much longer than ten minutes (*e.g.* Neidig, Grosser, and Hrovat, 1994), the white-light emission on active lower-mass M dwarfs can reach $> 10^{34}$ ergs and persist for hours (Hawley and Pettersen, 1991; Kowalski *et al.*, 2010, hereafter K10). More is known about the spectral shape of the white light during M dwarf flares (many spectrographs can easily obtain low-resolution, broad-wavelength spectra of stellar flares, but it is difficult to place a spectrograph slit over solar white-light kernels, which are intermittent and largely unpredictable), which have been studied using broadband colors (Hawley *et al.*, 2003; Zhilyaev *et al.*, 2007) and optical/NUV spectra (Hawley *et al.*, 2003; Hawley and Pettersen, 1991; Eason *et al.*, 1992; García-Alvarez *et al.*, 2002; Fuhrmeister *et al.*, 2008). In contrast to solar observations, the spectral shape during M dwarf flares suggests a hot blackbody with temperatures between $\approx 8500 - 11\,000$ K. Areal coverages of this component are typically $< 0.1\%$ of the visible stellar hemisphere, which implies a compact geometry like those observed in white light at the footpoints of flare arcades on the Sun. Although this blackbody component seems to be nearly ubiquitous during (large) stellar flares, it is not predicted even by the most recent one-dimensional radiative hydrodynamic (RHD) flare models (Allred *et al.*, 2006) produced with the RADYN code (Carlsson and Stein 1994, 1995, 1997).

The "megaflare" of 16 January 2009 on the dM4.5e star YZ CMi is one of the longest-lasting and most energetic flares observed on a low-mass single star. Low-resolution spectra (3350–9260 Å) were obtained in the flare's decay phase, which was elevated between 15 and 37 times the quiescent level and contained many secondary peaks. More than 160 spectra were obtained over 1.3 hours, and simultaneous U-band photometric observations of the entire flare event were provided by the New Mexico State University (NMSU) one-meter telescope. A detailed description of the observations and data reduction is given in K10.

In K10, two continuum components were necessary to fit the blue (3350–5500 Å) spectra: a hydrogen Balmer continuum (BaC) component as predicted by the RHD models of Allred *et al.* (2006) and a $T \approx 10\,000$ K blackbody component. An intriguing anti-correlation was found between the temporal evolution of these two components: the blackbody emission increased when the BaC decreased, and *vice versa* (see Figure 1d of K10). In this article, we revisit this anti-correlation and explain it using the phenomenological models of the secondary flare spectra from (Kowalski *et al.*, 2011a, hereafter K11). Finally, we show how each component of this flare might appear in the context of a solar-flare "arcade."

2. Anti-correlated Continuum Components and Continuum Veiling

The anti-correlation between the blackbody and BaC components (K10) can be understood qualitatively using Figure 1, where we show the spectral evolution of the total "flare-only" flux (denoted here as F_λ') during the rise phase of the secondary flare at $t \approx 130$ minutes. The spectra are color-coded to the nearest (in time) U-band measurement in the inset panel. At times prior to and near the beginning of the secondary flare (black, purple, and dark blue

Figure 1 A series of 16 flare spectra obtained prior to the onset and through the peak of a secondary flare. The quiescent spectrum from 24 November 2008 has been subtracted, as in K10. The U-band light curve (inset) is color-coded to the spectrum obtained closest in time. The best-fit blackbody curves to the black- and red-colored spectra are shown as the short-dashed ($T \approx 10\,400$ K) and long-dashed ($T \approx 13\,000$ K) lines, respectively.

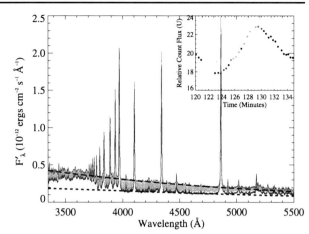

spectra), two distinct continuum components are clearly present in the spectra. The best-fit blackbody (short-dashed line) accounts for most continuum emission at $\lambda > 4000$ Å, whereas the BaC emission above the blackbody is conspicuous at $\lambda < 3750$ Å. During the rise and at the peak of the secondary flare (green, yellow, and red spectra), the BaC component seemingly disappears and the best-fit blackbody (long-dashed line) can fit the continuum shape throughout the entire wavelength range. K10 showed that the Hγ line flux exhibits an anti-correlated relation with the blackbody component. This effect is also present in Figure 1: in the red (secondary flare peak) spectrum, the continuum at $\lambda \approx 4200$ Å is highest, yet the peaks of the hydrogen Balmer lines are lowest.

The secondary flares at $t \approx 95$ minutes and $t \approx 130$ minutes are events during which the blackbody flux becomes stronger while the BaC flux becomes weaker. K10 quantified this as an increasing filling factor (areal coverage; percent of stellar disk) of the blackbody (with constant temperature, $T = 10\,000$ K) during the rise phase of each secondary flare. In Figure 1, we present an alternative interpretation. The blackbody curves (dashed lines) were fit to the spectra by allowing both the temperature and filling factor to vary. The best-fit blackbody temperatures and filling factors are $T \approx 10\,400$ K and $X_{BB} \approx 0.1\%$ (at $t = 122.9$ minutes; short-dashed line) and $T \approx 13\,000$ K and $X_{BB} \approx 0.1\%$ (at $t = 130$ minutes; long-dashed line). Strikingly, if both parameters are allowed to vary when fitting a blackbody function to these *total* flare spectra, the temperature increases by ≈ 2500 K while the filling factor remains approximately constant. Fitting a blackbody to the *total* flare spectrum (either by holding T constant, or by allowing X and T to vary) gives only the *average* properties of the entire flaring region at that time. We next show that these interpretations can be improved by isolating the newly formed flare emission.

K11 found that the isolated flare emission (denoted here as F_{λ}'') during the secondary flare's rise phase resembles the spectrum of a hot star, with the defining features being the hydrogen Balmer continuum and lines in *absorption* and a steeply rising continuum towards the blue at $\lambda > 4000$ Å. (Figure 1b of K11 shows that the new flare emission, obtained by subtracting the pre-secondary flare spectrum (average of three black and purple spectra around $t = 123.4$ minutes) from the average of two green-colored spectra around $t = 126.5$ minutes in Figure 1 of this article, is very similar to the spectrum of the A0 star Vega.) The observed anti-correlation between the continuum components in Figure 1 is a result of a "hot star spectrum" forming during the secondary flare. The hot star ("blackbody-like") spectrum causes an increase in the continuum at $\lambda \approx 4200$ Å by an amount, F_{4200}'', whereas an increase in the continuum at $\lambda \approx 3500$ Å occurs by only $\approx 0.6 \times F''_{4200}$. In other words, the flux in the continuum on both sides of the Balmer jump increases, but the continuum at $\lambda \approx 4200$ Å increases by a larger amount. Thus, the apparent decrease in the total amount of BaC in emission from $t = 123$ minutes to 130 minutes occurs as a result of "continuum veiling" (similar to the continuum veiling observed for accreting T Tauri stars – *e.g.*, Hartigan *et al.*, 1989; Hessman and Guenther, 1997; Herczeg and Hillenbrand, 2008).

3. Combining Continuum Components Using Phenomenological Hot Spot Models

In K11, the F''_{λ} emission was modeled phenomenologically with the static radiative transfer code RH (Uitenbroek, 2001), as a temperature bump ("hot spot") with peak temperature $T = 20\,000$ K, placed near the photosphere (below the temperature minimum) of the quiescent M-dwarf atmosphere. Here, we use a sum of individual hot spots and the RHD model spectrum (hereafter called RHDF11) of Allred *et al.* (2006) to model the *total* flare emission $[F'_{\lambda}]$ at two times during the megaflare on YZ CMi. Figure 2 shows flare spectra from Figure 1 averaged around $t = 123.4$ minutes ($F0'$; gray) and at $t = 126.5$ minutes ($F1'$; black). These are the spectra corresponding to times immediately before and nearly halfway up the rise phase of the secondary flare, respectively (*i.e.*, the same two spectra presented in Figure 1a of K11 but with the quiescent level subtracted).

To model $F0'$, we add the RHDF11 spectrum and a hot spot (HS1) with $T_{max} = 12\,000$ K (keeping the other parameters the same as described in K11) with a ratio of filling factors of $10:1$ and $X_{RHDF11} = 1.2\%$, as found in K10. The total model spectrum is shown as the light blue (short dashes) curve in Figure 2. As in K11, we model the secondary flare as a hot spot (HS2) with $T_{max} = 20\,000$ K. Adding HS2 to RHDF11 and HS1 gives the red (long dashes) spectrum in Figure 2. The areal coverage of HS2 is 0.46 as large as the areal coverage of HS1. These model spectra match the observed continuum levels at all wavelengths in the figure. Moreover, the continuum-veiling/anti-correlation effect is readily apparent: the height of the Balmer jump at $\lambda = 3646$ Å decreases from black (no hot spots) to blue (one hot spot) to red (two hot spots). The model fluxes are slightly lower than the observed fluxes at $\lambda < 3750$ Å. At these wavelengths, there is a forest of metallic lines (*e.g.*, Fe I, Fe II) that are blended in our low-resolution spectra; our model is satisfactory in matching the *underlying* level which is likely closer to the actual level of the BaC. Note

Figure 2 Flare spectra at $t = 123.4$ minutes and 126.5 minutes from Figure 1 are shown in gray and black, respectively. The composite model spectra are shown in blue, short dashes $(F_{\lambda,RHDF11} \times X_{RHDF11} + F_{\lambda,HS1} \times X_{HS1})$ and red, long dashes $(F_{\lambda,RHDF11} \times X_{RHDF11} + F_{\lambda,HS1} \times X_{HS1} + F_{\lambda,HS2} \times X_{HS2})$. The RHDF11 spectrum is shown as the thin black line. The continuum-veiling effect is apparent from the different heights of the Balmer discontinuity at $\lambda = 3646$ Å.

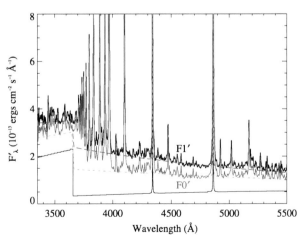

that in K10, we intentionally used only the BaC from the RHDF11 spectrum to model the continuum, whereas the RHDF11 predictions for the Paschen continuum and photospheric-backwarming components are included in the total fluxes in this work.

The origin of the hot spots is unknown, as they are not predicted by self-consistent RHD models that employ a solar-type nonthermal electron heating function (Allred *et al.*, 2006). In the proposed continuum model, we use the fewest number of components necessary to fit the overall shape and reproduce the anti-correlation. However, HS1 may represent a sum of individual decaying hot spots from previously heated flare regions (see below). We are working to produce a grid of phenomenological models that will be used to constrain the column mass of the hot spots, the detailed temperature evolution, and the uniqueness of the continuum fit. RHD models (with RADYN) of the gradual phase are also forthcoming and will more accurately represent the BaC and photospheric backwarming in the decay phase.

4. The Solar Analogy

Figure 3 shows how a spatially resolved observation of the YZ CMi megaflare might appear. We use the continuum components and filling factors to make analogies to several flare structures and phenomena observed in large solar flare arcades. The complex morphology of the U-band light curve leads us to speculate that the YZ CMi megaflare involved a large arcade, or several large arcades of flare loops. The main features of Figure 3 are the following:

BaC (yellow ribbons): Throughout the spectral observations (72 minutes $< t <$ 149 minutes), the hydrogen BaC and lines were highly elevated and decreasing (likely from the initial flare peaks in the U-band light curve), implying that this emission had originated from the footpoints of a previously heated magnetic arcade in the flaring chromosphere.

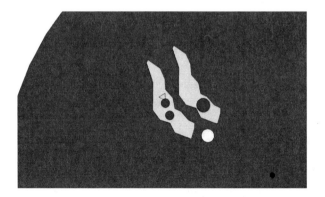

Figure 3 Graphic with continuum components (areas from $t \approx 126.5$ minutes to scale) as they might appear in a spatially resolved observation. The BaC emitting region (yellow) resembles a two-ribbon structure (shown here as symmetric for simplicity) in a thin layer of the heated mid-to-upper chromosphere of previously reconnected magnetic loops. HS1 is shown as a collection of previously formed hot spots (purple), and HS2 is the proposed newly formed continuum emitting region (white). The triangle indicates the assumed location of the initial flare peak, which generated a disturbance in the lower atmosphere that propagated into the surrounding active region and triggered the hot spots. The black circle helps orient the reader to the center of the star, which has a radius $0.3 R_\odot$. The flare region is placed at an arbitrary location on the surface. Several aspects of this cartoon were inspired by observations of solar-flare arcades, such as Fletcher and Hudson (2001).

These may manifest as a complex of flare ribbons, as is commonly observed in Hα during solar flares; *e.g.*, Rust and Hegwer (1975), Berlicki *et al.* (2004), Balasubramaniam *et al.* (2010). In some solar flares (*e.g.*, Neidig, 1983), the BaC appears to have a spatial morphology that is more compact than an extended Hα ribbon. Spectroscopic observations are needed to compare the plasma properties and conditions of BaC and Balmer line emitting ribbons and kernels.

HS1 (purple spots): Immediately prior to the secondary flare beginning at $t \approx 123$ minutes, a series of phenomenological hot spots (HS1) appear near the photosphere. These hot spots were formed during the previous secondary flares (*i.e.*, at $t \approx 65$ minutes, ≈ 95 minutes; see K10), and they are emitting from a total source size that is $\approx 1/10$ as large as the area of the chromospheric flare region (*e.g.*, Hα ribbons). The spectra of these hot spots have the hydrogen BaC and lines in absorption. They might be similar to the compact white-light kernels during solar flares, as in Wang, Fang, and Ding (2007), Fletcher *et al.* (2007), Jess *et al.* (2008), or they may be similar to Ellerman-bomb phenomena (see Section 4.1). Also emitting from the photosphere is a larger region heated from chromospheric (BaC) backwarming; we assume that the size of this backwarmed region is the same size as the flaring chromosphere.

HS2 (white spot): The secondary flare at $t \approx 130$ minutes is the result of the formation of a new hot spot (HS2), hotter and smaller than HS1 but at the same column mass. At this time, we see a sudden decline in the BaC flux. When all of these components are unresolved, as in our stellar spectra, continuum veiling produces the observed anti-correlation. We have placed HS2 assuming it was triggered by a disturbance induced by the huge initial flare peak at $t \approx 28$ minutes. The time evolution of the Hγ and BaC fluxes in Figure 1d of K10 indicates an apparent lack of new hydrogen Balmer-line emitting regions during the secondary U-band peaks (we cannot definitively determine whether there is a newly formed hydrogen Balmer emitting (chromospheric ribbon) component cospatial with the hot spot because the observations are unresolved). Therefore, the disturbance likely propagated through the lower atmosphere, below the height of hydrogen Balmer-line formation (upper chromosphere; J. Allred, private communication 2010). Using a range of sound speeds in the lower atmosphere for the speed of the disturbance ($\approx 5 - 10$ km s^{-1}), we find that HS2 is located at a distance that is approximately $30 - 60$ Mm ($R_{\text{YZ CMi}} \approx 200$ Mm) from the site of the initial flare event.[1]

The composite graphical model is preliminary (see Section 5) and requires comparison to other complex flare events on dMe stars but especially to solar flares where we can spatially resolve each continuum component. Our group is currently working to obtain solar-flare data that can be used to test the YZ CMi flare model using the Dunn Solar Telescope (DST) with the Rapid Oscillations in the Solar Atmosphere (ROSA) imager and employing custom continuum filters (Jess *et al.*, 2010b; Kowalski *et al.*, 2011b).

4.1. Are the Secondary Flares Stellar Ellerman Bombs?

Ellerman bombs are transient, compact phenomena observed near evolving or emerging magnetic fields in solar active regions (Ellerman, 1917; Severny, 1968; Georgoulis *et al.*,

---

[1] The white-light and hard X-ray footpoints have been observed to propagate along the polarity inversion line during large solar flares such as the famous 14 July 2000 flare (Fletcher and Hudson, 2001; Kosovichev and Zharkova, 2001; Qiu *et al.*, 2000), but the spatial location of these kernels appears to change much faster, $\approx 170 - 200$ km s^{-1}.

2002, and references therein). A typical signature of Ellerman bombs is emission in the wings and absorption in the core of Hα relative to the nearby plage intensity (Koval and Severny, 1970; Bruzek, 1972; Fang *et al.*, 2006). The temporal evolution properties include mean lifetimes of ≈ 10 − 20 minutes and fine-structure variations (Kurokawa *et al.*, 1982; Qiu *et al.*, 2000). In contrast to typical white-light flares, Ellerman bombs have symmetric light curves with similar rise and decay times (Payne, 1993; Qiu *et al.*, 2000; Jess *et al.*, 2010a). The Ellerman-bomb mechanism is not fully understood but has been attributed to magnetic reconnection in the low chromosphere (*e.g.*, Georgoulis *et al.*, 2002).

The secondary flares during the YZ CMi megaflare exhibit several similarities to Ellerman-bomb phenomena on the Sun. Ellerman bombs have also been modeled phenomenologically as temperature bumps at or below the solar temperature minimum region (Fang *et al.*, 2006; Berlicki, Heinzel, and Avrett, 2010). The secondary YZ CMi flares have longer rise times (≈ two – five minutes) and are much more symmetric about the peak compared to other white-light flares with similar total energy on YZ CMi ($\Delta t_{rise} \approx 0.5$–1.8 minutes; Moffett, 1974; van den Oord *et al.*, 1996). The absorption features of the blackbody-like continuum component are similar to the line-center absorption observed in Hα and Ca II during Ellerman bombs; unfortunately, our observations do not have sufficient spectral resolution to separate line-center and wing profiles. The preliminary finding (Section 4) that the blackbody-like continuum component does not contain hydrogen line emission may be consistent with magnetic reconnection taking place in the low atmosphere. However, in contrast to solar Ellerman bombs, which have been observed as a microflare trigger (Jess *et al.*, 2010a), the secondary flares are possibly a *consequence* of the enormous YZ CMi flare peak event.

5. Summary and Future Work

The time-resolved continuum data obtained during a megaflare on the dM4.5e star YZ CMi demonstrate the power of broad-wavelength coverage, low-resolution spectra, which are unfortunately not currently available in the optical/near-UV region for solar flares. In this manuscript, we show that the blackbody-like component (hot star-like emission) of the white-light continuum dominates the spectra during the secondary flares while the BaC (and Balmer lines) become less important; this observed anti-correlation is explained as continuum veiling. We combine the phenomenological models of the blackbody-like component from K11 and the Allred *et al.* (2006) RHD model spectrum of the BaC to reproduce the total flare emission at two times during the flare. The filling factors for the decaying BaC (plus backwarming) component, a previously heated hot spot component, and a newly heated hot spot component are in the ratio of ≈ 10 : 1 : 0.5. These areal coverages allow us to compare each component with observed solar flare structures in large eruptive flare arcades. Although we generally assume that stellar-flare phenomena are simply "scaled-up" versions of solar-flare phenomena, one should not exclude the possibility that stellar flares might have fundamental differences in the white-light continuum, as the energies and time scales of dMe flares can be orders of magnitude larger than solar flares. New solar-flare observations are needed to test the existence of the blackbody-like component and to better understand the properties of the BaC, which could be fully "unveiled" using spatially resolved solar observations.

We provide evidence that the blackbody-like component has several similar properties to solar Ellerman bombs. A few solar flares have exhibited the spectral features (Svestka, 1963) and velocity characteristics (Kosovichev and Zharkova, 2001) of Ellerman bombs.

Additional intensity-calibrated continuum measurements of Ellerman bombs and white-light flare kernels on the Sun, such as with DST/ROSA, would help illuminate the differences between these events and provide a comparison to stellar spectra of typical white-light flares and megaflare-size events.

Several aspects of the phenomenological models presented in this work are being improved. In addition to modeling hydrogen with more levels and including metallic transitions and molecular species, a more accurate consideration of charge balance is underway. The correct treatment of charge balance in a modified atmosphere is complicated by non-LTE ionization, but a new version of RH has been provided by H. Uitenbroek to account for this; the authors are currently working on a new suite of hot-spot models.

Acknowledgements   This work was first presented at "The Origin, Evolution, and Diagnosis of Solar Flare Magnetic Fields and Plasmas: Honoring the Contributions of Dick Canfield," a conference that took place from 9 – 11 August 2010 at NCAR/HAO in Boulder, CO. AFK thanks the organizers of this conference for generous travel assistance and acknowledges support from NSF grant AST 0807205. We gratefully thank H. Uitenbroek for many useful discussions and for our use of the RH code; P. Heinzel, M. Varady, and D. Jess for illuminating conversations about Ellerman bombs; and J. Allred for allowing us to use the detailed output of his RADYN flare models. We also acknowledge Google for our use of its SketchUp application. This work is based on observations obtained with the Apache Point Observatory 3.5 m Telescope, which is owned and operated by the Astrophysical Research Consortium.

References

Allred, J.C., Hawley, S.L., Abbett, W.P., Carlsson, M.: 2006, *Astrophys. J.* **644**, 484. doi:10.1086/503314.

Balasubramaniam, K.S., Cliver, E.W., Pevtsov, A., Temmer, M., Henry, T.W., Hudson, H.S., Imada, S., Ling, A.G., Moore, R.L., Muhr, N., Neidig, D.F., Petrie, G.J.D., Veronig, A.M., Vršnak, B., White, S.M.: 2010, *Astrophys. J.* **723**, 587. doi:10.1088/0004-637X/723/1/587.

Berlicki, A., Heinzel, P., Avrett, E.H.: 2010, *Mem. Soc. Astron. Ital.* **81**, 646.

Berlicki, A., Schmieder, B., Vilmer, N., Aulanier, G., Del Zanna, G.: 2004, *Astron. Astrophys.* **423**, 1119. doi:10.1051/0004-6361:20040259.

Bruzek, A.: 1972, *Solar Phys.* **26**, 94. doi:10.1007/BF00155110.

Carlsson, M., Stein, R.F.: 1994, In: Carlsson, M. (ed.) *Chromospheric Dynamics, University Oslo*, 47.

Carlsson, M., Stein, R.F.: 1995, *Astrophys. J. Lett.* **440**, 29. doi:10.1086/187753.

Carlsson, M., Stein, R.F.: 1997, *Astrophys. J.* **481**, 500. doi:10.1086/304043.

Donati-Falchi, A., Smaldone, L.A., Falciani, R.: 1984, *Astron. Astrophys.* **131**, 256.

Eason, E.L.E., Giampapa, M.S., Radick, R.R., Worden, S.P., Hege, E.K.: 1992, *Astron. J.* **104**, 1161. doi:10.1086/116305.

Ellerman, F.: 1917, *Astrophys. J.* **46**, 298. doi:10.1086/142366.

Fang, C., Tang, Y.H., Xu, Z., Ding, M.D., Chen, P.F.: 2006, *Astrophys. J.* **643**, 1325. doi:10.1086/501342.

Fletcher, L., Hudson, H.: 2001, *Solar Phys.* **204**, 69. doi:10.1023/A:1014275821318.

Fletcher, L., Hannah, I.G., Hudson, H.S., Metcalf, T.R.: 2007, *Astrophys. J.* **656**, 1187. doi:10.1086/510446.

Fuhrmeister, B., Liefke, C., Schmitt, J.H.M.M., Reiners, A.: 2008, *Astron. Astrophys.* **487**, 293. doi:10.1051/0004-6361:200809379.

García-Alvarez, D., Jevremović, D., Doyle, J.G., Butler, C.J.: 2002, *Astron. Astrophys.* **383**, 548. doi:10.1051/0004-6361:20011743.

Georgoulis, M.K., Rust, D.M., Bernasconi, P.N., Schmieder, B.: 2002, *Astrophys. J.* **575**, 506. doi:10.1086/341195.

Hartigan, P., Hartmann, L., Kenyon, S., Hewett, R., Stauffer, J.: 1989, *Astrophys. J. Suppl. Ser.* **70**, 899. doi:10.1086/191361.

Hawley, S.L., Pettersen, B.R.: 1991, *Astrophys. J.* **378**, 725. doi:10.1086/170474.

Hawley, S.L., Allred, J.C., Johns-Krull, C.M., Fisher, G.H., Abbett, W.P., Alekseev, I., Avgoloupis, S.I., Deustua, S.E., Gunn, A., Seiradakis, J.H., Sirk, M.M., Valenti, J.A.: 2003, *Astrophys. J.* **597**, 535. doi:10.1086/378351.

Herczeg, G.J., Hillenbrand, L.A.: 2008, *Astrophys. J.* **681**, 594. doi:10.1086/586728.

Hessman, F.V., Guenther, E.W.: 1997, *Astron. Astrophys.* **321**, 497.

Hiei, E.: 1982, *Solar Phys.* **80**, 113. doi:10.1007/BF00153427.

Hudson, H.S., Acton, L.W., Hirayama, T., Uchida, Y.: 1992, *Publ. Astron. Soc. Japan* **44**, 77.

Jess, D.B., Mathioudakis, M., Crockett, P.J., Keenan, F.P.: 2008, *Astrophys. J.* **688**, 119. doi:10.1086/595588.

Jess, D.B., Mathioudakis, M., Browning, P.K., Crockett, P.J., Keenan, F.P.: 2010a, *Astrophys. J. Lett.* **712**, L111. doi:10.1088/2041-8205/712/1/L111.

Jess, D.B., Mathioudakis, M., Christian, D.J., Keenan, F.P., Ryans, R.S.I., Crockett, P.J.: 2010b, *Solar Phys.* **261**, 363. doi:10.1007/s11207-009-9500-0.

Kosovichev, A.G., Zharkova, V.V.: 2001, *Astrophys. J. Lett.* **550**, L105. doi:10.1086/319484.

Koval, A.N., Severny, A.B.: 1970, *Solar Phys.* **11**, 276. doi:10.1007/BF00155226.

Kowalski, A.F., Hawley, S.L., Holtzman, J.A., Wisniewski, J.P., Hilton, E.J.: 2010, *Astrophys. J. Lett.* **714**, L98. doi:10.1088/2041-8205/714/1/L98.

Kowalski, A.F., Hawley, S.L., Holtzman, J.A., Wisniewski, J.P., Hilton, E.J.: 2011a, In: Choudhary, D.P., Strassmeier, K.G. (eds.) *Physics of Sun and Star Spots, Proc. Int. Astron. Un.* **6**, Symposium S273, 261–264. doi:10.1017/S1743921311015341

Kowalski, A.F., Mathioudakis, M., Keys, P.H., Hawley, S.L.: 2011b, *Solar Phys.*, in preparation.

Kurokawa, H., Kawaguchi, I., Funakoshi, Y., Nakai, Y.: 1982, *Solar Phys.* **79**, 77. doi:10.1007/BF00146974.

Machado, M.E., Rust, D.M.: 1974, *Solar Phys.* **38**, 499. doi:10.1007/BF00155084.

Moffett, T.J.: 1974, *Astrophys. J. Suppl. Ser.* **29**, 1. doi:10.1086/190330.

Neidig, D.F.: 1983, *Solar Phys.* **85**, 285. doi:10.1007/BF00148655.

Neidig, D.F., Kane, S.R.: 1993, *Solar Phys.* **143**, 201. doi:10.1007/BF00619106.

Neidig, D.F., Grosser, H., Hrovat, M.: 1994, *Solar Phys.* **155**, 199. doi:10.1007/BF00670740.

Payne, T.E.W.: 1993, A multiwavelength study of solar Ellerman bombs. Ph.D. thesis, New Mexico State University.

Qiu, J., Ding, M.D., Wang, H., Denker, C., Goode, P.R.: 2000, *Astrophys. J. Lett.* **544**, L157. doi:10.1086/317310.

Qiu, J., Liu, W., Hill, N., Kazachenko, M.: 2000, *Astrophys. J.* **725**, 319. doi:10.1088/0004-637X/725/1/319.

Rust, D.M., Hegwer, F.: 1975, *Solar Phys.* **40**, 141. doi:10.1007/BF00183158.

Severny, A.B.: 1968, In: Öhman, Y. (ed.) *Mass Motions in Solar Flares and Related Phenomena*, Almqvist and Wiksell, Stockholm, 71.

Svestka, Z.: 1963, *Bull. Astron. Inst. Czech.* **14**, 234.

Uitenbroek, H.: 2001, *Astrophys. J.* **557**, 389. doi:10.1086/321659.

van den Oord, G.H.J., Doyle, J.G., Rodono, M., Gary, D.E., Henry, G.W., Byrne, P.B., Linsky, J.L., Haisch, B.M., Pagano, I., Leto, G.: 1996, *Astron. Astrophys.* **310**, 908.

Wang, L., Fang, C., Ding M.-D.: 2007, *Chin. J. Astron. Astrophys.* **7**, 721. doi:10.1088/1009-9271/7/5/13.

Zhilyaev, B.E., Romanyuk, Y.O., Svyatogorov, O.A., Verlyuk, I.A., Kaminsky, B., Andreev, M., Sergeev, A.V., Gershberg, R.E., Lovkaya, M.N., Avgoloupis, S.J., Seiradakis, J.H., Contadakis, M.E., Antov, A.P., Konstantinova-Antova, R.K., Bogdanovski, R.: 2007, *Astron. Astrophys.* **465**, 235. doi:10.1051/0004-6361:20065632.

Solar Phys (2012) 277:31–44
DOI 10.1007/s11207-011-9823-5

SOLAR FLARE MAGNETIC FIELDS AND PLASMAS

Optical-to-Radio Continua in Solar Flares

P. Heinzel · E.H. Avrett

Received: 17 January 2011 / Accepted: 28 May 2011 / Published online: 11 August 2011
© Springer Science+Business Media B.V. 2011

Abstract Spectral continua observed during solar flares may contain information about both thermal and non-thermal heating mechanisms. Using two semi-empirical flare models F2 and FLA, we synthesize the thermal continua from optical to mm–radio domains and compare their intensities with quiet-Sun values computed from a recent model C7. In this way, the far-infrared and sub-mm/mm continua are studied for the first time, and we present our results as a benchmark for further modeling and for planning new observations, especially with the ALMA instrument. Finally, we demonstrate how these continua are formed and show a close correspondence between their brightness temperature and the kinetic-temperature structure of the flaring atmosphere.

Keywords Solar flares · Spectral continua · Diagnostics

1. Introduction

Spectral diagnostics of solar flares, ranging from high-energy spectra up to the radio domain, provide us with important information in two respects: Firstly, the thermal structure of the flaring atmosphere can be determined in different locations, generally depending on time, and this is directly related to heating mechanisms at work. Secondly, many attempts have been undertaken to extract, from the observed spectral data, information about non-thermal processes that are supposed to be due to highly energetic electron or proton beams interacting with the flare atmosphere at various heights. Unfortunately, thermal and non-thermal

Solar Flare Magnetic Fields and Plasmas
Guest Editors: Y. Fan and G.H. Fisher

P. Heinzel (✉)
Astronomical Institute, Academy of Sciences, 251 65 Ondřejov, The Czech Republic
e-mail: pheinzel@asu.cas.cz

E.H. Avrett
Harvard-Smithsonian Center for Astrophysics, 60 Garden Street, Cambridge, MA, USA
e-mail: avrett@cfa.harvard.edu

effects on spectra are in most cases mixed and thus it is rather difficult to reliably separate information about particle beams and other processes. Various spectral lines can serve as typical diagnostics, since their formation is affected by thermal as well as non-thermal processes – *e.g.*, Kašparová *et al.* (2009a) demonstrating the formation of hydrogen Balmer lines in a beam-heated chromospheric flare.

There are fewer studies of various spectral continua, to some extent as a result of the lack of appropriate observations. However, new possibilities arise thanks to highly sophisticated instrumentation currently in use or being constructed, for both ground and space observations. Hydrogen Balmer and Paschen continua are still not well understood in solar and stellar flares (*e.g.* the problem of the so-called backwarming) and new observations are planned using the ROSA instrument (Jess *et al.*, 2010) and *SOlar Robotic Telescope* (SORT) at the Ondřejov Observatory. Understanding the white-light continua is also challenging. In the near-infrared (NIR) spectral range, a strong enhancement of emission at 1.56 microns was recently detected (Xu *et al.*, 2006) and this is rather difficult to explain (see recent radiation-hydrodynamical simulations of Cheng, Ding, and Carlsson, 2010). Even more unexplored continuum regions are the far-infrared (FIR) and sub-mm/mm (SMM) radio continua. This is again dictated by various observational difficulties. However, the latter continua have a great diagnostic potential for both thermal and non-thermal processes: Thermal FIR and radio continua are directly related to the thermal structure of the chromosphere and transition region as we will demonstrate in this article. On the other hand, the non-thermal processes manifest themselves through the high-frequency part of the spectrum from ultra-relativistic electrons and/or positrons, *i.e.* the synchrotron emission which is added on top of the thermal continuum (Vial *et al.*, 2007). In the FIR domain, the relative importance of these processes was discussed by Ohki and Hudson (1975).

To better understand the formation of the FIR and radio continua, we start by modeling their thermal component. For this we exploit two types of so-called semi-empirical flare models and synthesize the flare continua. We also demonstrate at which atmospheric locations these continua are formed depending on the model. Implications for analysis of newly planned observations are then drawn.

2. Observations

Here we briefly summarize the new tools aimed at observing the FIR–radio continua. In fact our initial motivation to study FIR continua was related to the planned space experiment *SMall Explorer for Solar Eruption* (SMESE) which was designed to carry a FIR spectrometer *Detection of Eruptive Solar Infrared Emission* (DESIR) onboard; see Vial *et al.* (2007). The FIR continuum window was between 35 and 250 microns, the range which is not accessible from ground and thus so far unexplored. Although the DESIR experiment has not so far advanced, its FIR range represents an important diagnostic window. On the other hand, it was interesting to realize that the same window – and even wider – has been selected for the ESA *Herschel* space FIR telescope (http://www.esa.int/SPECIALS/Herschel/). This provides an opportunity to observe the FIR continua on flare stars, and *Herschel* observing time has been already allocated for this purpose. Going to longer wavelengths, the radio SMM continua have been occasionally observed (Trottet *et al.*, 2002, see also recent review by Krucker *et al.*, 2011). New wide-band (0.3 – 9.6 mm) observations with high spatial resolution are planned with the ESO-ALMA interferometer (Karlický *et al.*, 2011). Therefore, it is highly desirable to carry out modeling of FIR–radio continua in flares, both solar and stellar.

3. Semi-empirical Models

Semi-empirical models of solar flares have been constructed in order to derive the thermal structure of the flaring atmosphere, consistently with observations in various spectral lines and continua. The models are usually based on hydrostatic and statistical equilibria. Assuming a given temperature structure, the full non-LTE transfer problem is solved numerically to obtain the synthetic flare spectrum. This is then compared with the observed spectral features and the temperature structure is properly adjusted to get better agreement. This process is iterated until a reasonable match is achieved. This method, which avoids the explicit treatment of the energy-balance problem (which is highly uncertain due to poorly understood heating–cooling mechanisms) allows us to gain reasonable information on the atmospheric temperature structure, provided that the atmosphere is not evolving on very short time scales (see discussion in Carlsson and Stein, 1995). In highly dynamical cases, the meaning of a semi-empirical temperature is questionable, particularly when the modeling is based on UV data where the time averaging is strongly non-linear (Carlsson and Stein, 1995). However, in the FIR–radio domain, the situation is less critical. It is also important to distinguish between the impulsive and gradual phases of flares. The latter are more appropriate for semi-empirical modeling.

In this article we use as a benchmark the widely known semi-empirical flare models F2 and FLA. The model F2 represents the atmospheric model constructed for flares of medium importance, and its Hα line emission agrees with typical flare observations (Machado *et al.*, 1980). Two other models, F1 and F3, correspond to weak and strong flares, respectively. Slight modifications to these models were proposed by Avrett, Machado, and Kurucz (1986), where a catalogue of theoretical line profiles was presented for a variety of spectral lines. The model is in hydrostatic equilibrium and its temperature structure is shown in Figure 1. Similarly, model FLA (Mauas, Machado, and Avrett, 1990) was constructed to derive the temperature structure of a white-light flare. Its photospheric temperature (Figure 1) is enhanced to fit the white-light observations of Mauas (1990). Both F2 and FLA models were constructed using the PANDORA radiative-transfer code (Avrett and Loeser, 2003). The same code is used to synthesize the continuum spectra in this article. Note that the height dependence of the temperature used in this study is the same as in original models. However, since we neglect the microturbulent velocities and since the ionization structure is computed by the current (largely improved) version of PANDORA, the column-mass scale is somewhat different from the original one.

Finally, to compare with quiet-Sun conditions, we use the most recent semi-empirical model of the quiet solar atmosphere, based on SOHO/SUMER spectra (Avrett and Loeser, 2008). This model is called C7 and its temperature structure is also shown in Figure 1. Around the temperature minimum and higher-up, both F2 and FLA models differ strongly from C7. At greater depths, *i.e.* at photospheric layers, where the optical continuum is typically formed, the difference is small or negligible. However this region is somewhat tricky. In Figure 1 the model F2 reaches C7 at these depths, while FLA temperatures are somewhat higher. This is consistent with the optical-continuum observations: models F1, F2, and F3 are based on flare observations not exhibiting the white-light continuum enhancement. On the other hand, FLA is the model for a white-light flare. In this article we have adjusted the model F2 to C7 at photospheric depths, while the original F2 model of Machado *et al.* (1980) was adjusted to the quiet-Sun photosphere of the model VAL3C (Vernazza, Avrett, and Loeser, 1981). We show this original adjustment by the thin dashed line in Figure 1 and the spectral consequences are mentioned below. Other differences between C7 and VAL3C are shown in Figure 2 of Avrett (2007), but these are not relevant to our flare modeling.

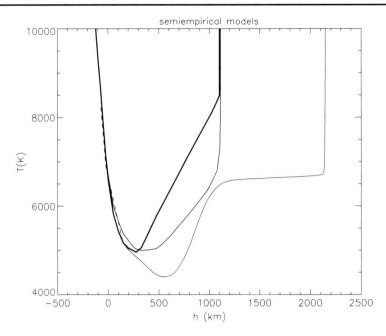

Figure 1 Temperature structure of the three semi-empirical models. C7: thin line, FLA: medium line, F2: thick line. A thin dashed line indicates the photospheric part of the former VAL3C quiet-Sun model, to be compared with slightly modified C7 photosphere. The original F2 flare model used this VAL3C photosphere.

Also note that the transition-region part of C7 is based on energy balance with ambipolar diffusion (Fontenla, Avrett, and Loeser, 1990).

4. Formation of FIR–Radio Continua

The problem of the formation of continuum intensities at the wavelengths under consideration is largely simplified because the continuum source function is given by the local Planck function (*i.e.* for free–free processes detailed below), $S = B_\nu(T)$, where T is the kinetic temperature. However, the continua are not formed under simplified LTE conditions, because the chromospheric opacities depend on neutral hydrogen, electron, and proton densities and these must be computed under fully non-LTE conditions. In a simple static, time-independent case, these particle densities follow from statistical equilibrium solutions, where hydrogen is the dominant species. In dynamical, time-dependent cases, the situation is even more complex (see Kašparová *et al.*, 2009b).

In the chromosphere, the dominant source of opacity is the hydrogen free–free continuum for which the absorption coefficient at frequency ν is given as (Rybicki and Lightman, 1979)

$$\kappa_\nu(\mathrm{H}) = 3.7 \times 10^8 T^{-1/2} n_e n_p \nu^{-3} g_{\mathrm{ff}}, \tag{1}$$

where n_e and n_p are the electron and proton densities, respectively, T is the kinetic temperature, and $g_{\mathrm{ff}} \approx 1$ is the Gaunt factor. Around the temperature minimum region, H$^-$ free–free opacity also plays an important role. This absorption is according to Kurucz (1970)

$$\kappa_\nu(\mathrm{H}^-) = \frac{n_e n_{\mathrm{H}}}{\nu}\left(1.3727 \times 10^{-25} + \left(4.3748 \times 10^{-10} - 2.5993 \times 10^{-7}/T\right)/\nu\right), \tag{2}$$

where n_H is the neutral hydrogen density. The total free–free absorption coefficient corrected for stimulated emission is then

$$\kappa_\nu = \left[\kappa_\nu(H) + \kappa_\nu(H^-)\right]\left(1 - e^{-h\nu/kT}\right), \tag{3}$$

where h and k are the Planck and Boltzmann constants, respectively. Note the significant importance of the stimulated emission term in the FIR–radio domain. The emergent synthetic intensity I_ν is obtained as

$$I_\nu = \int B_\nu(T)e^{-\tau_\nu}\, d\tau_\nu = \int \eta_\nu e^{-\tau_\nu}\, dh, \quad \eta_\nu = \kappa_\nu B_\nu, \quad d\tau_\nu = -\kappa_\nu\, dh, \tag{4}$$

where $B_\nu(T)$ is the Planck source function, η_ν the emission coefficient, τ_ν the optical depth, and h is the geometrical height in the atmosphere. We can also write

$$I_\nu = \int CF\, dh, \quad CF = \eta_\nu e^{-\tau_\nu}, \tag{5}$$

where we have introduced the wavelength-dependent contribution function $CF = dI_\nu/dh$ (Carlsson, 1998; Avrett and Loeser, 2008). Its behavior depends on the kinetic-temperature and non-LTE ionization structure of the flaring atmosphere and will be discussed in the next section. For simplicity of exposition we have omitted the μ-dependence of the specific intensity (μ is the cosine of the viewing angle).

In the FIR–radio domain, I_ν and B_ν are directly proportional to the brightness temperature $[T_b]$ and to the local plasma (kinetic) temperature $[T]$, respectively (e.g., Rybicki and Lightman, 1979)

$$I_\nu = \frac{2\nu^2 k}{c^2} T_b, \quad B_\nu = \frac{2\nu^2 k}{c^2} T, \tag{6}$$

where c is the speed of light. Using this Rayleigh–Jeans law, Equation (4) can be written as

$$T_b = \int T e^{-\tau_\nu}\, d\tau_\nu = \int T e^{-\tau_\nu} \kappa_\nu\, dh. \tag{7}$$

Finally, assuming a unique dependence of T on height, we can transform this equation to the form

$$T_b = \int CF'\, dT, \tag{8}$$

where

$$CF' = \frac{T e^{-\tau_\nu} \kappa_\nu}{f(T)}, \quad f(T) = \frac{dT}{dh}. \tag{9}$$

This shows how the observed wavelength-dependent brightness temperature is synthesized depending on the height distribution of the plasma temperature which affects the shape of the contribution function CF'. We will use this to interpret our numerical results.

5. Synthetic Spectra and Formation Depths

The principal aim of this article is to compute the synthetic spectra of selected optical-to-radio continua, which correspond to semi-empirical flare models used in the literature. These

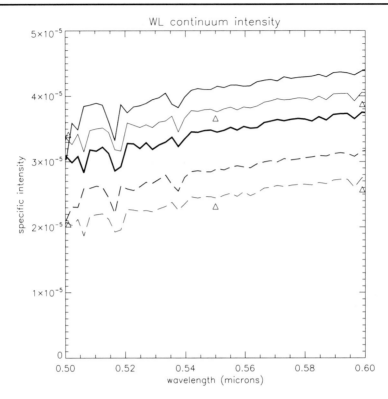

Figure 2 Computed WL (optical) continuum in the range 0.5–0.6 microns. C7: thin line, FLA: medium line, F2: thick line. Full lines: disk center, dashed lines for FLA and C7: $\mu = 0.4$ (the limb-darkening effect is evident). Triangles are the observed data from Allen (1973). Intensities are in cgs units [erg sec^{-1} cm^{-2} sr^{-1} Hz^{-1}]. Note the effect of many overlapping lines at shorter wavelengths.

spectra can be considered as benchmarks for further studies of the continuum formation and simultaneously as a reference for planning new observations and for their subsequent analysis.

We start first with presenting a portion of the white-light (WL) continuum in Figure 2. The quiet-Sun WL continuum was computed with the latest model C7 and at three selected wavelengths (triangles) was compared with the photospheric observations, both for disk center and for $\mu = 0.4$. The WL continuum of the model FLA is somewhat enhanced as expected and this is consistent with the modeling of Mauas, Machado, and Avrett (1990) which is based on WL-flare observations of Mauas (1990). Finally, model F2 deserves some attention. As mentioned above and shown in our Figure 1, at photospheric levels the original model F2 of Machado *et al.* (1980) approached the photosphere of VAL3C and this produces WL continuum shown in Figure 2. It is somewhat below the observed values. Replacing the photospheric part of F2 by the C7 temperature structure would lead to WL-continuum intensity values similar to those for C7 in Figure 2. Therefore, F2 with a C7 photosphere is more consistent with WL observations, while FLA photosphere is based on enhanced WL emission in flare observations and thus has nothing to do with model C7, except for deeper photospheric layers.

Motivated by recent observations of the near-infrared (NIR) continuum during a flare (Xu *et al.*, 2006), we have also computed this portion of the continuum spectrum, between 1

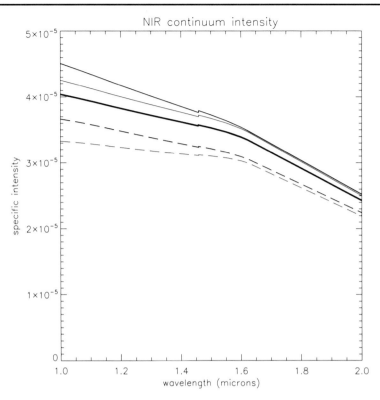

Figure 3 Computed NIR continuum in the range 1 – 2 microns. C7: thin line, FLA: medium line, F2: thick line. Full lines: disk center, dashed lines for FLA and C7: $\mu = 0.4$ (the limb-darkening effect is still evident). Intensities are in cgs units [erg sec^{-1} cm^{-2} sr^{-1} Hz^{-1}].

and 2 microns (Figure 3). The models used and line annotation are the same as in the case of the WL continuum. The behavior is similar to WL continuum, we see a slight enhancement in the case of a WL flare (model FLA). But none of these semi-empirical models, even the rather strong F2 flare, is capable of producing relatively strong emission at 1.56 microns as the observations indicate. Note that even very recent time-dependent simulations of the electron-beam heating have not led to better agreement (Cheng, Ding, and Carlsson, 2010). Therefore, the continuum emission at this NIR window is not yet understood and further such observations are required, complemented by non-LTE modeling. Note finally that the synthesis of WL and NIR continua requires rather complex computations of various opacities which was done by the PANDORA code. Namely in the short-wavelength part of the WL continuum, overlapping of many lines is evident (see Figure 2).

For FIR-to-radio continua, we have computed the specific intensities for all three models in the wavelength range between 30 microns and 10 mm. This covers the FIR and SMM ranges mentioned in Section 2. In particular, the latter are now highly relevant with respect to planned ALMA observations (0.3 – 9.6 mm). In Figure 4a we see an enhanced FIR continuum as compared to quiet-Sun and this enhancement significantly increases at longer wavelengths. At 10 mm the intensity for the F2 flare model is almost two orders of magnitude larger than for the C7 model. The WL flare model is somewhere in between. The quiet-Sun is supposed to be observed by ALMA at a wavelength range covering the chromospheric layers (Loukitcheva *et al.*, 2004; Avrett and Loeser, 2008;

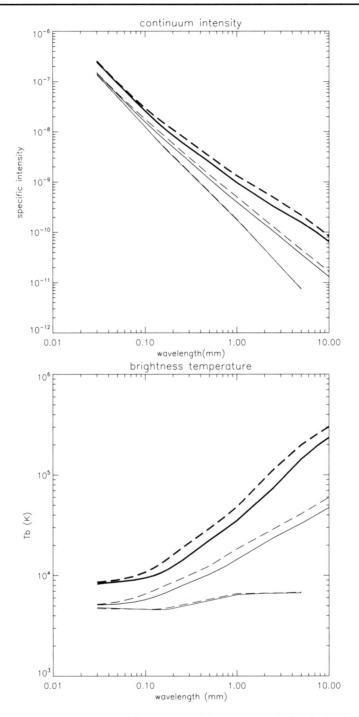

Figure 4 (a) Specific intensities computed for the three models in the FIR–radio domain. C7: thin line, FLA: medium line, F2: thick line. Full lines: disk center, dashed lines for $\mu = 0.4$. Note the limb-brightening effect for flare models. Intensities are in cgs units [erg sec^{-1} cm^{-2} sr^{-1} Hz^{-1}]. (b) Brightness temperatures T_b for same cases.

Karlický *et al.*, 2011). Therefore, the enhanced flare brightness should be also well detected by ALMA. Moreover, on top of this, there are the non-thermal continua which are due to synchrotron radiation of electron beams. In Figure 4b we show the same plot but the specific intensities are converted to equivalent brightness temperatures. In the FIR–radio domain, these T_b values should correspond to plasma kinetic temperatures at corresponding depths of the continuum formation. This depends on the wavelength. From Figure 4b we see that for the FIR range, the quiet-Sun continuum is formed around the temperature minimum region, while temperatures higher in the chromosphere are well reflected in the ALMA range. On the other hand, for both flares FLA and F2, the FIR–radio continua correspond to the enhanced chromospheric and transition-region temperatures. The brightness temperature of the F2 continuum is much larger than that of C7, in the ALMA wavelength range. Note also an effect of limb brightening which indicates that we observe the chromospheric layers assuming the source function equal to the local Planck function.

To demonstrate the formation of FIR–radio continua more rigorously, we plot in Figures 5 and 6 the contours of the contribution function $CF(h)$, which, for a given wavelength, depends on the atmospheric height h. Plotted CFs are normalized such that their integral in Equation (5) is equal to one. At the shortest FIR wavelengths, the FLA continuum is formed around the temperature minimum, while at longer FIR wavelengths it is formed already in the chromosphere. A similar plot for the F2 model (Figure 6) shows that the FIR continuum in the flaring chromosphere is fully formed above 1000 km. In both the FLA and F2 cases, the radio (ALMA) continuum is then formed within a very narrow range of atmospheric depths, as indicated in zoomed portions of the plots, see Figures 5b and 6b, respectively. Note that for the C7 model one can find the CF functions in the range of FIR–radio continua in Avrett and Loeser (2008) (their Figure 2).

It is also important to understand how these contribution functions transform the plasma kinetic temperatures into the synthetic brightness temperatures. In other words, how precise a FIR–radio instrument can be in measuring the plasma temperature, *i.e.* how good a thermometer it is. This is shown in Figure 7, which was constructed in the following way: CF in Figures 5a and 6a were first transformed to T-scale, using the gradients dT/dh from Figure 1. Apart from a normalization factor, this gives the dependence of CF' defined in Equation (9) on wavelength and temperature T. By integrating this CF' over T (see Equation (8)), we obtain T_b at given wavelength. This is the same as integrating CF in Figures 5 and 6 over h, which would give us, for a given wavelength, the specific intensity. Finally, we just replaced the wavelengths at which the brightness temperature is synthesized by T_b itself, using the dependence presented in Figure 4b. Figure 7 thus represents a 2D plot of CF', where both axes have transformed to temperatures. It has a meaningful interpretation only along the horizontal direction, like the contribution functions in Figures 5 and 6. The maxima of CF' indicate a good correlation between T_b and T, but the breadth along the T-axis clearly shows the contributions to the observed T_b from other depths with different temperatures. Only in the case of FLA, the temperatures around and above the temperature minimum are not so well reproduced, apparently due to more complex transfer effects.

For a given model, we can easily find to which atmospheric depth the continuum formation temperature is related. Using Equations (1) and (2), we have computed the optical depths for the FIR–radio domain and found that the CF typically peaks at layers where the continuum becomes optically thin at a given wavelength ($\tau \approx 1$ moves from the chromosphere at FIR wavelengths to the transition region at SMM). So these are roughly the depths of the continuum formation, although there is sometimes a fainter optically thin component formed above. But the latter does not seem to affect the brightness temperatures substantially, as our Figure 7 indicates. Note that we did not consider here still higher layers with

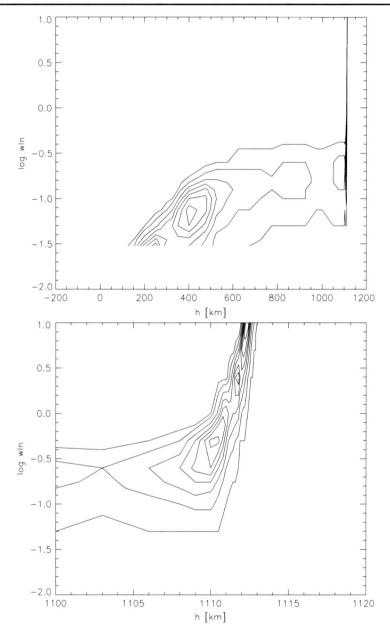

Figure 5 Contribution functions for the FLA model. The contours represent the percentage of CF, the step is 3%. wln stands for wavelengths in mm. The lower panel shows a zoom of the transition region.

temperatures above 10^6 K, *i.e.* the region where the soft X-ray emission is formed in flare loops. This region can also contribute to the continuum brightness, depending on the actual hot-plasma source (see Ohki and Hudson, 1975; Krucker *et al.*, 2011).

However, the information contained in Figure 7 just says that what we measure is a height variation of the kinetic temperature, so we can decide how much it varies, compared

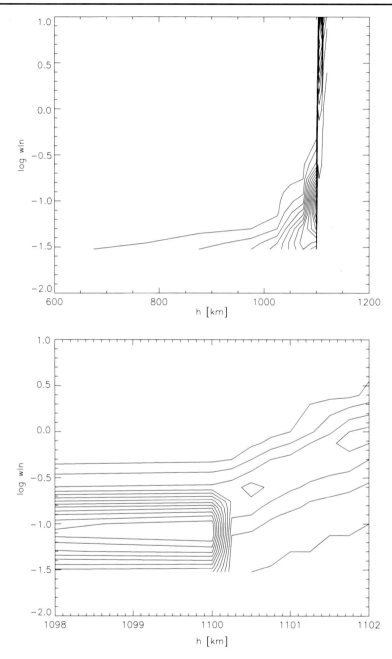

Figure 6 Contribution functions for the F2 model. The contours represent the percentage of CF, the step is 10% which indicates in general more concentrated CF than in case of the FLA model. wln stands for wavelengths in mm. The lower panel shows a zoom of the transition region.

with, e.g., the quiet-Sun one, but we cannot easily relate these temperatures to atmospheric heights. This requires more sophisticated modeling of the type discussed in this article. Note also that the FIR–radio continuum brightness has no direct relation to WL-continuum

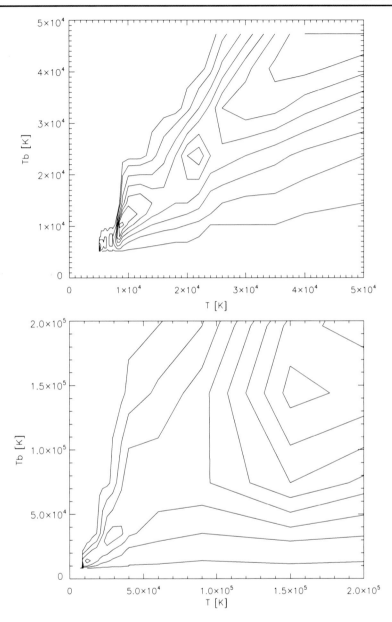

Figure 7 Brightness temperature T_b vs. the plasma kinetic temperature T, for models FLA (a) and F2 (b). Construction of these plots is explained in the text.

intensity because both continua are formed at quite different atmospheric regions (*i.e.* the upper chromosphere *vs.* regions below the temperature minimum). Therefore, the enhanced FIR–radio continuum in case of FLA model is not necessarily the signature of a WL flare.

Finally, we should note that under flare conditions, where the electron densities are significantly enhanced relative to quiet Sun, the radiation transport in the radio mm domain may be affected by a wave reflection at frequencies close to the plasma frequency (Kašparová *et al.*, 2009b). To check this quantitatively, we use the relation (*e.g.* Aschwanden, 2004)

$$\nu_p = 8979\sqrt{n_e}, \tag{10}$$

where ν_p is the plasma frequency and n_e the electron density. For enhanced flare densities around 10^{13} cm^{-3}, found within a chromospheric condensation of the model F2 (roughly for T between 10^4 and 2×10^4 K), the plasma frequencies lie around 2.8×10^{10} Hz, which corresponds to wavelengths close to 10 mm. This may be a problem for interpretation of future ALMA observations of solar flares, but will not be critical for studies of the quiet chromosphere.

6. Conclusions and Future Prospects

In this article we have mainly focused on FIR-to-radio (mm) thermal continua as formed during solar flares. Our results represent benchmarks for planning new observations and for further modeling. They are specifically relevant to those flare evolutionary stages when the non-thermal beams are unimportant or not present at all, so that first no additional synchrotron component is expected and, second, the atmospheric ionization structure is purely thermal. Such a situation is typical for later phases of the flare evolution, *e.g.* for a gradual phase. For initial impulsive phases, the non-thermal processes of excitation and ionization should be included (see Kašparová and Heinzel (2002) for the case of model F1) which will modify the neutral hydrogen, electron, and proton densities entering Equations (1) and (2). This can be modeled in the same way as by Kašparová and Heinzel (2002), and we plan to perform such simulations in a subsequent article. Finally, further detailed study should account for the (fast) temporal evolution of the flaring atmosphere during the impulsive phase. FIR–radio continua for such radiation-hydrodynamical flare models were synthesized for the first time by Kašparová *et al.* (2009b), in a similar way as Loukitcheva *et al.* (2004) did for the quiet chromosphere, and this work will also continue. Finally, we plan to add the synchrotron component as a diagnostics of precipitating electron beams. Such a synchrotron component, due to electrons in a strong magnetic field, was recently reported by Cristiani *et al.* (2008), based on sub-mm observations. But, surprisingly, the authors claim that the free–free thermal emission does not contribute to the observed brightness. This contradicts our present results and thus further quantitative analysis of new data will be extremely important to better constrain the flare models.

Acknowledgements   The authors thank H. Hudson, G. Trottet, J.-P. Raulin, and an anonymous referee for useful comments. P.H. was the member of the Team 165 of L. Fletcher and J. Kašparová at ISSI – Bern, where the problems related to flare continua have been extensively discussed. He also acknowledges support from grant P209/10/1680 (Grant Agency of the Czech Republic) and appreciates the kind hospitality of the Center for Astrophysics, Cambridge (USA), where part of this work was done.

References

Allen, C.W.: 1973, *Astrophysical Quantities*, Athlone Press, London.
Aschwanden, M.J.: 2004, *Physics of the Solar Corona*, Springer, Berlin.
Avrett, E.H.: 2007. In: Heinzel, P., Dorotovič, I., Rutten, R.J. (eds.) *The Physics of Chromospheric Plasmas* CS-368, Astron. Soc. Pac., San Francisco, 81.
Avrett, E.H., Loeser, R.: 2003. In: Piskunov, N., Weiss, W.W., Gray, D.F. (eds.) *Modeling of Stellar Atmospheres, IAU Symp.* **210**, Astron. Soc. Pac., San Francisco, A21.
Avrett, E.H., Loeser, R.: 2008, *Astrophys. J. Suppl.* **175**, 229.

Avrett, E.H., Machado, M.E., Kurucz, R.L.: 1986. In: Neidig, D.F. (ed.) *The Lower Atmosphere of Solar Flares*, NSO, Sunspot, 216.

Carlsson, M.: 1998. In: Vial, J.-C., Bocchialini, K., Boumier, P. (eds.) *Space Solar Physics, Lecture Notes in Physics* **507**, Springer, Berlin, 163.

Carlsson, M., Stein, R.F.: 1995, *Astrophys. J. Lett.* **440**, L29.

Cheng, J.X., Ding, M.D., Carlsson, M.: 2010, *Astrophys. J.* **711**, 185.

Cristiani, G., Giménez de Castro, C.G., Mandrini, C.H., Machado, M.E., de Benedetto e Silva, I., Kaufmann, P., Rovira, M.G.: 2008, *Astron. Astrophys.* **492**, 215.

Fontenla, J.M., Avrett, E.H., Loeser, R.: 1990, *Astrophys. J.* **355**, 700.

Jess, D.B., Mathioudakis, M., Christian, D.J., Keenan, F.P., Ryans, R.S.I., Crockett, P.J.: 2010, *Solar Phys.* **261**, 363.

Karlický, M., Bárta, M., Dabrowski, B., Heinzel, P.: 2011, *Solar Phys.* **268**, 165.

Kašparová, J., Heinzel, P.: 2002, *Astron. Astrophys.* **382**, 688.

Kašparová, J., Varady, M., Heinzel, P., Karlický, M., Moravec, Z.: 2009a, *Astron. Astrophys.* **499**, 923.

Kašparová, J., Heinzel, P., Karlický, M., Moravec, Z., Varady, M.: 2009b, *Cent. Eur. Astrophys. Bull.* **33**, 309.

Krucker, S., Bastian, T., Giménez de Castro, C.G., Hales, A.S., Hudson, H.S., Kašparová, J., Klein, K.-L., Kretzschmar, M., Lüthi, T., Mackinnon, A., Pohjolainen, S., Trottet, G., White, S.M.: 2011, *Astron. Astrophys. Rev.*, submitted.

Kurucz, R.L.: 1970, SAO Special Rep. **309**.

Loukitcheva, M.A., Solanki, S.K., Carlsson, M., Stein, R.F.: 2004, *Astron. Astrophys.* **419**, 747.

Machado, M.E., Avrett, E.H., Vernazza, J.E., Noyes, R.W.: 1980, *Astrophys. J.* **242**, 336.

Mauas, P.J.D.: 1990, *Astrophys. J. Suppl.* **74**, 609.

Mauas, P.J.D., Machado, M.E., Avrett, E.H.: 1990, *Astrophys. J.* **360**, 715.

Ohki, K., Hudson, H.S.: 1975, *Solar Phys.* **43**, 405.

Rybicki, G.B., Lightman, A.P.: 1979, *Radiative Processes in Astrophysics*, Wiley-Interscience, New York.

Trottet, G., Raulin, J.-P., Kaufmann, P., Siarkowski, M., Klein, K.-L., Gary, D.E.: 2002, *Astron. Astrophys.* **381**, 694.

Vernazza, J.E., Avrett, E.H., Loeser, R.: 1981, *Astrophys. J. Suppl.* **45**, 635.

Vial, J.-C., Auchere, F., Chang, J., Fang, C., Gan, W.Q., Klein, K.-L., Prado, J.-Y., Trottet, G., Wang, C., Yan, Y.H.: 2007, *Adv. Space Res.* **40**, 1787.

Xu, Y., Cao, W., Liu, Ch., Yang, G., Jing, J., Denker, C., Emslie, A.G., Wang, H.: 2006, *Astrophys. J.* **641**, 1210.

Solar Phys (2012) 277:45–57
DOI 10.1007/s11207-011-9796-4

The Evolution of Sunspot Magnetic Fields Associated with a Solar Flare

Sophie A. Murray · D. Shaun Bloomfield ·
Peter T. Gallagher

Received: 20 December 2010 / Accepted: 12 May 2011 / Published online: 12 July 2011
© Springer Science+Business Media B.V. 2011

Abstract Solar flares occur due to the sudden release of energy stored in active-region magnetic fields. To date, the precursors to flaring are still not fully understood, although there is evidence that flaring is related to changes in the topology or complexity of an active-region's magnetic field. Here, the evolution of the magnetic field in active region NOAA 10953 was examined using *Hinode*/SOT-SP data over a period of 12 hours leading up to and after a GOES B1.0 flare. A number of magnetic-field properties and low-order aspects of magnetic-field topology were extracted from two flux regions that exhibited increased Ca II H emission during the flare. Pre-flare increases in vertical field strength, vertical current density, and inclination angle of $\approx 8°$ toward the vertical were observed in flux elements surrounding the primary sunspot. The vertical field strength and current density subsequently decreased in the post-flare state, with the inclination becoming more horizontal by $\approx 7°$. This behavior of the field vector may provide a physical basis for future flare-forecasting efforts.

Keywords Active regions, magnetic fields · Flares, relation to magnetic field · Magnetic fields, photosphere · Sunspots, magnetic fields

1. Introduction

Active regions in the solar atmosphere have complex magnetic fields that emerge from subsurface layers to form loops that extend into the corona. When active regions un-

Solar Flare Magnetic Fields and Plasmas
Guest Editors: Y. Fan and G.H. Fisher

S.A. Murray (✉) · D.S. Bloomfield · P.T. Gallagher
Astrophysics Research Group, School of Physics, Trinity College Dublin, Dublin 2, Ireland
e-mail: somurray@tcd.ie

D.S. Bloomfield
e-mail: shaun.bloomfield@tcd.ie

P.T. Gallagher
e-mail: peter.gallagher@tcd.ie

dergo external forcing, the system may destabilise and produce a solar flare, where energy stored in sunspot magnetic fields is suddenly released as energetic particles and radiation across the entire solar spectrum (Rust, 1992; Conlon *et al.*, 2008). The initial impulsive phase of the flare is generally believed to be driven by magnetic reconnection, which leads to a change in the topology of the magnetic field, and energy stored in the field is released, accelerating coronal particles (Aschwanden, 2005). The storage of magnetic energy in active regions is indicated by the degree of non-potentiality of sunspot magnetic fields (Régnier and Priest, 2007). The processes leading up to reconnection and energy release are still not fully understood, and studying the links between solar flares and topology changes in active-region magnetic fields is an important step in understanding the pre-flare configuration and the process of energy release (Hewett *et al.*, 2008; Conlon *et al.*, 2010).

Many early theoretical studies suggested a link between both the emergence of new flux and the shearing and twisting of field lines with the flare trigger mechanism (Rust *et al.*, 1994). Shearing is taken to mean that the field is aligned almost parallel to the neutral line rather than perpendicular to it, as would be observed in a potential field (Schmieder *et al.*, 1996). Tanaka (1986) depicts a possible evolution of large-scale fields in a flare, with an ensemble of sheared fields containing large currents and a filament located above the neutral line in the pre-flare state. Canfield, Leka, and Wülser (1991) explored the importance of strong currents further, finding that sites of significant energetic-electron precipitation into the chromosphere were at the edges of regions of strong vertical current rather than within them. Metcalf *et al.* (1994) and Li *et al.* (1997) subsequently found that flares do not necessarily coincide spatially with the locations of strong vertical current. More sophisticated flare models were later developed, *e.g.* Antiochos (1998) described a "breakout" model for large eruptive flares, with newly emerged, highly sheared field held down by an overlying unsheared field. Field topology studies have been used to place constraints on theoretical models; *e.g.* Mandrini (2006) reviewed a number of flaring active-region topologies, finding that magnetic reconnection can occur in a greater variety of magnetic configurations than traditionally thought. The reader is referred to the reviews of Priest and Forbes (2002) and Schrijver (2009), and references therein, for more recent developments in eruptive event models.

Numerous observational studies have confirmed the importance of emergence and shearing to flare phenomena. Zirin and Wang (1993) investigated flux emergence and sunspot group motions, which resulted in complicated flow patterns leading to flaring. Wang *et al.* (1994) used vector magnetograms to observe magnetic shear in five X-class solar flares; in all cases increasing along a substantial portion of the magnetic neutral line. They suggested flux emergence being key to eruption, but the increase in shear persisted much longer after the flare rather than decreasing as per model predictions. No definitive theoretical explanation was given. Recent evidence has furthered the idea that emerging-flux regions and magnetic helicity are crucial to the pre-flare state (*e.g.*, Liu and Zhang, 2001; Wang *et al.*, 2002; Chandra *et al.*, 2009), where magnetic helicity is a measure of magnetic topological complexity, *e.g.* twists and kinks of field lines (Canfield and Pevtsov, 1998). Line-of-sight (LOS) magnetic-field observations have shown that photospheric fields change rapidly during large solar flares (Sudol and Harvey, 2005; Petrie and Sudol, 2009). Other studies use improved extrapolation techniques to analyse the topology further, increasing our understanding of eruptions in the solar corona (Régnier and Canfield, 2006; Georgoulis and LaBonte, 2007). Observing active-region magnetic fields around the time of flaring can be very beneficial, as magnetic-field properties have been found to be viable flare-forecasting tools (Gallagher, Moon, and Wang, 2002). However, the LOS magnetic field alone cannot provide complete information on the changing magnetic field.

High spatial resolution observations of the solar magnetic-field vector can now provide more in-depth information on the true 3D topological complexities. In this article we use spectropolarimetric measurements from the *Hinode* spacecraft (Tsuneta *et al.*, 2008) to examine how sunspot magnetic fields evolve leading up to and after flare activity. In particular, differences in the magnetic-field vector between pre- and post-flare states are examined in the vicinity of a chromospheric flare brightening. Studying the evolution of the magnetic field before the flare with these improved observations could outline some new flare precursors that may of be use in flare forecasting, perhaps in terms of how soon a flare could be expected after certain conditions are met. Any changes observed after the flare compared to the pre-flare conditions should also give insight into how a flare might occur from this kind of region, testing the validity of currently proposed changes in magnetic topology during solar flares (*e.g.*, Pevtsov, Canfield, and Zirin, 1996). In Section 2 we briefly discuss the observations and data-analysis techniques used. Section 3 presents the main results, in particular the changes in vertical and horizontal field in Section 3.1, field orientation in Section 3.2, and derived low-order 3D magnetic properties in Section 3.3. Finally, our main conclusions and directions for future work are outlined in Section 4.

2. Observations and Data Analysis

Active region NOAA 10953 (http://www.solarmonitor.org/region.php?date=20070426& region=10953) crossed the solar disk from 26 April 2007 to 9 May 2007. Previous studies of this region have found evidence of twisting; *e.g.* Canou and Amari (2010) examined the magnetic structure of the region on 30 April 2007. Their reconstructed magnetic configurations exhibited twisted flux ropes along the southern part of the neutral line, similar to observations by Okamoto *et al.* (2009) of twisted flux ropes emerging from below the photosphere. Here, we use observations of the main sunspot on 29 April 2007 recorded by the *Solar Optical Telescope* (SOT: Suematsu *et al.*, 2008) onboard *Hinode*. Table 1 lists the scan start and end times and pointing information. The simple-structured active region consisted of a negative-polarity leading sunspot and opposite-polarity trailing plage, with an "S-shaped" filament visible over this time. In addition, this region was the source of a low-magnitude GOES B1.0 solar flare: beginning at 10:34 UT; peaking at 10:37 UT; ending at 10:40 UT.

Four scans from the SOT spectropolarimeter (SP: Kosugi *et al.*, 2007) were used, with a scan duration of ≈ 32 minutes each. The temporal scan coverage was a critical reason for choosing this event, *i.e.* three scans before the flare and one immediately after (Table 1). Using multiple scans prior to the flare enables the non-flare-related evolution of the magnetic-field properties to be analysed in detail, with changes over the flare able to be compared to this background evolution. No other flares occurred during the entire time period of observation, avoiding the contamination of any of the scans.

SOT-SP recorded the Stokes I, Q, U, and V profiles of the Fe I 6301.5 Å and 6302.5 Å lines simultaneously through a 0.16″ × 164″ slit. The Stokes spectral profiles were recorded with a spectral sampling of 21.5 mÅ, a field-of-view (FOV) of 164″ × 164″ (512 × 512 pixels), and an exposure time of 3.2 seconds per slit position (fast map mode). The raw SOT-SP data were calibrated using sp_prep.pro from the *Hinode*/SOT tree within the IDL SolarSoft library (Freeland and Handy, 1998), which makes two passes through the data. The first determines the thermal shifts in the spectral dimension (in both offset and dispersion) across successive slit positions. The second pass corrects these thermal variations and merges the two orthogonal polarization states.

Table 1 Summary of SOT-SP scan times on 29 April 2007.

Scan Number	Begin Time (UT)	End Time (UT)	Center of FOV (Solar X, Solar Y)
1	00:17	00:49	$-549''$, $-99''$
2	03:30	04:02	$-525''$, $-98''$
3	08:00	08:32	$-491''$, $-96''$
Flare[a]	10:34	10:40	$-476''$, $-150''$
4	11:27	11:59	$-464''$, $-95''$

[a]Flare location corresponds to reconstructed RHESSI image peak.

The resulting Stokes I, Q, U, and V profiles were inverted using the *He-Line Information Extractor* (HELIx^{+}: Lagg *et al.*, 2004) in order to derive the magnetic-field vector. HELIx^{+} fits the observed Stokes profiles with synthetic ones obtained from an analytic solution of the Unno–Rachkovsky (Unno, 1956) equations in a Milne–Eddington atmosphere. The model atmosphere used in fitting the observed profiles consisted of one magnetic component with a local straylight component included. Optimal atmospheric parameters are obtained using PIKAIA, a genetic algorithm-based general-purpose optimization subroutine (Charbonneau, 1995). A total polarization threshold of $\approx 3 \times 10^{-3}\ I_{c}$ (*i.e.* units of continuum intensity) was chosen, such that regions with values below this were not inverted.

The AMBIG routine (Leka, Barnes, and Crouch, 2009), which is an updated form of the Minimum Energy Algorithm (Metcalf, 1994), was used to remove the 180° ambiguity in the LOS azimuthal angle. This procedure was chosen over other routines as it scored highly in the Metcalf *et al.* (2006) and Leka *et al.* (2009) reviews on methods for resolving solar ambiguity angles. The routine simultaneously minimises the magnetic-field divergence $[\nabla \cdot \mathbf{B}]$ and vertical electric-current density $[J_{z}]$ for pixels above a certain noise threshold in transverse-field strength. In this work we take a value of 150 G, whereby pixels with values below this level are determined using an iterative acute-angle-to-nearest-neighbors method (Canfield *et al.*, 1993).

The resulting LOS inversion results were converted to the solar-surface normal reference frame using the method outlined by Gary and Hagyard (1990). The orthogonal magnetic-field components in the observers (*i.e.* image, superscript "i") frame and solar-surface normal (*i.e.* heliographic, superscript "h") frame are related by

$$B_{x}^{h} = a_{11}B_{x}^{i} + a_{12}B_{y}^{i} + a_{13}B_{z}^{i},$$
$$B_{y}^{h} = a_{21}B_{x}^{i} + a_{22}B_{y}^{i} + a_{23}B_{z}^{i}, \qquad (1)$$
$$B_{z}^{h} = a_{31}B_{x}^{i} + a_{32}B_{y}^{i} + a_{33}B_{z}^{i},$$

where coefficients a_{ij} are defined in Equation (1) of Gary and Hagyard (1990). In the image frame, B_{z}^{i} is the component along the LOS, and (B_{x}^{i}, B_{y}^{i}) define the plane of the image. In the heliographic frame, B_{z}^{h} is the component normal to the solar surface, and (B_{x}^{h}, B_{y}^{h}) lie in the plane tangent to the solar surface at the center of the FOV. In terms of the field vector, $B_{x}^{h} = |\mathbf{B}|\sin(\gamma)\cos(\phi)$, $B_{y}^{h} = |\mathbf{B}|\sin(\gamma)\sin(\phi)$, and $B_{z}^{h} = |\mathbf{B}|\cos(\gamma)$. Here, $|\mathbf{B}|$ is the absolute magnetic field strength, γ is the inclination angle from the solar normal direction, and ϕ is the azimuthal angle in the (B_{x}^{h}, B_{y}^{h}) plane measured counter-clockwise from solar west.

The scans were taken $\approx$ three – four hours apart, so it was necessary to correct for changes in scan pointing. To solve this, all scans were differentially rotated and their continuum intensity co-aligned to that of the third scan. Examples of observations from the third scan (*i.e.* immediately preceding the flare) are shown in Figure 1, including *Hinode*/SOT-SP continuum intensity (Figure 1(a)) and resulting magnetic field parameters from the HELIx$^+$ code after disambiguation and transformation to the solar-normal reference frame: absolute magnetic-field strength (Figure 1(c)); inclination angle with azimuthal-angle vectors overlayed (Figure 1(d)); vertical field strength, B_z^h (Figure 1(e)); horizontal field strength, $B_{hor}^h = [(B_x^h)^2 + (B_y^h)^2]^{1/2}$ (Figure 1(f)).

SOT *Broadband Filter Imager* Ca II H line images (3968 Å) were also obtained close to the flare peak time, with a FOV of $108'' \times 108''$ (1024×1024 pixels^2). Figure 1(b) shows a Ca II H image at the time of the third scan, as well as contours of significant brightening at the time of the flare peak at 10:37 UT (1250 DN) overlaid on all other images. The brightening seems to be mostly located along the neutral line dividing the sunspot and plage regions in the East of the scan. The location containing the most significant chromospheric flare brightening is found to the South East (SE) of the main sunspot, located near the trailing plage neutral line. A $35'' \times 40''$ box was chosen from this region for analysis. The sub-region was divided into two specific regions of interest, ROI 1 and ROI 2, defined by thresholding the signed field magnitude (*i.e.* |**B**| times -1 or $+1$ for fields pointing in or out of the solar surface, respectively). ROI 1 was thresholded at -800 G and ROI 2 at -1000 G. Both of these regions are small flux elements of the same polarity as the main sunspot, and are located SE of the main spot. They both lie close to the neutral line with the positive-polarity plage (see Figure 1(e)). These two ROIs will be the focus of the magnetic field parameters studied.

3. Results

Figure 2 shows the temporal evolution of the magnetic field in the chosen sub-region over the four scans. ROI 1 fragments significantly from the first to the third pre-flare scans, and almost completely disappears after the flare. ROI 2 also fragments, but changes less than ROI 1. The chromospheric flare brightenings are located over and North West (NW) of ROI 1, and directly over ROI 2.

The parameters depicted in Figure 2 were separately analysed in detail for both ROIs. The median and standard deviation of the values were extracted from all pixels within a ROI contour in each individual scan. Median values were used rather than other averaging methods due to their ease of interpretation and relative insensitivity to outlying values. The structure of the field was investigated in different ways: the vector field components (Section 3.1); the field-orientation angles (Section 3.2); signatures of magnetic non-potentiality (Section 3.3).

3.1. Vector Field Components

Changes in ROI median values of the field magnitude, vertical field, and horizontal field were calculated in each scan (*i.e.* values from all pixels in the thresholded contours of a ROI). Figure 3 depicts time lines of these ROI median values, with vertical bars representing the ROI standard deviation and horizontal bars depicting the scan duration. The magnetic-field strength in Figure 3(a) varies little over all the scans within 1-σ errors, with only a slight decrease in the second scan for ROI 1. The horizontal-field strength, given in Figure 3(b), shows only a slightly decreasing trend over the scans. The main source of interest here comes from the vertical-field strength.

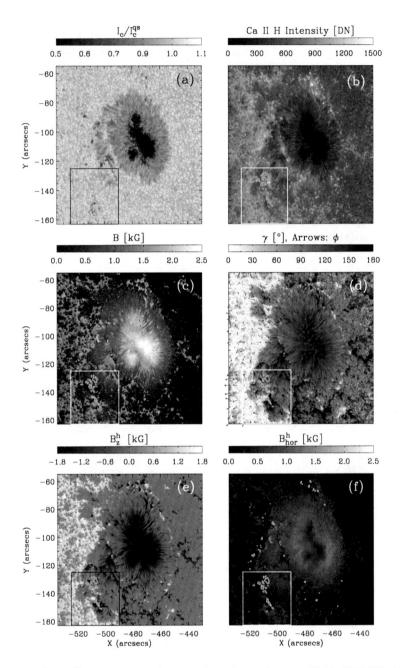

Figure 1 108″ × 108″ FOV images showing the active-region pre-flare state (08:00 – 08:32 UT): (a) continuum intensity; (b) Ca II H intensity (08:16 UT); (c) absolute field strength; (d) inclination angle, with transverse magnetic-field vectors overlaid as arrows (magenta); (e) vertical-field strength; (f) horizontal-field strength. Green contours in all panels outline the significant Ca II H flare brightening (at the 1250 DN level) observed at 10:37 UT. The sub-region selected for further analysis in Figure 2 is indicated by the box in all panels.

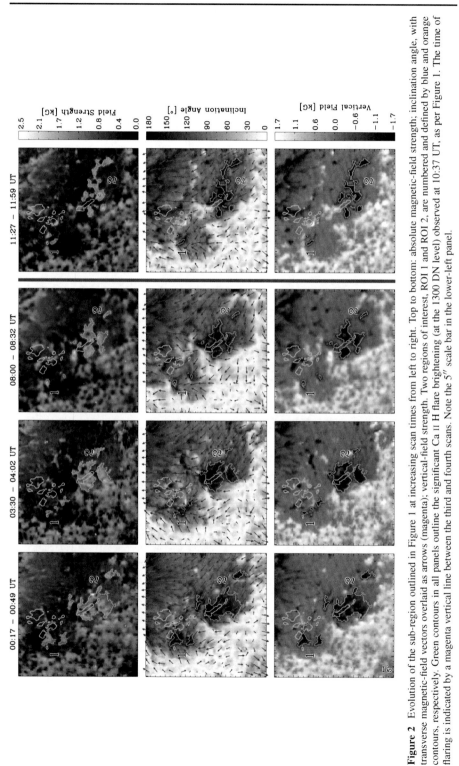

Figure 2 Evolution of the sub-region outlined in Figure 1 at increasing scan times from left to right. Top to bottom: absolute magnetic-field strength; inclination angle, with transverse magnetic-field vectors overlaid as arrows (magenta); vertical-field strength. Two regions of interest, ROI 1 and ROI 2, are numbered and defined by blue and orange contours, respectively. Green contours in all panels outline the significant Ca II H flare brightening (at the 1300 DN level) observed at 10:37 UT, as per Figure 1. The time of flaring is indicated by a magenta vertical line between the third and fourth scans. Note the 5″ scale bar in the lower-left panel.

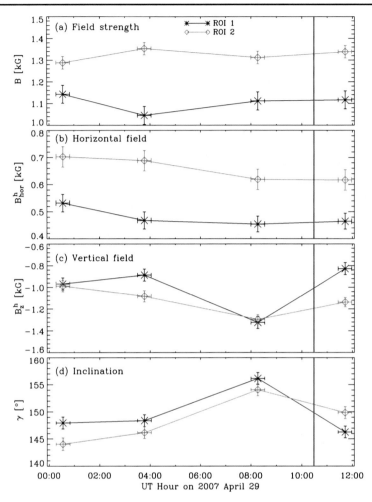

Figure 3 Temporal variation in the median values of: (a) absolute magnetic-field strength; (b) horizontal-field strength; (c) vertical-field strength; (d) inclination angle. Values for ROI 1 are plotted with black asterisks, and ROI 2 with blue diamonds. Vertical bars indicate the standard deviation of the property within the ROI, while horizontal bars delimit the scan duration. The vertical line between the third and fourth scans marks the flare peak time. With values of inclination being beyond $90°$, increasing values indicate the field becoming more vertical.

The vertical-field median value also marginally changes within the spread of ROI values between the first two scans, as can be seen in Figure 3(c). However, substantial variations are found between both the second and the third scans, as well as the third and fourth scans. An increase in vertical-field magnitude is found between the second and third scans, increasing by ≈ 440 G for ROI 1 and ≈ 210 G for ROI 2. After the flare (*i.e.* some time between the third and fourth scans) B_z^h decreases by ≈ 500 G for ROI 1 and ≈ 160 G for ROI 2. It is likely that the changes prior to the flare are linked to the energy storage mechanism in the ROIs, while the changes over the course of the flare are due to the energy release. However, it is unclear from the median-field magnitude measurements how the field structure is changing before and after the flare. Thus, field orientation was investigated further.

3.2. Field Orientation

The median inclination angle was also extracted from both ROIs and is included in Figure 3(d). A similar trend in inclination evolution is seen to the vertical-field evolution. Again no changes of significance are found between the first two scans, with large changes observed between the second and third scans and after the flare. An increase in inclination is found in the third scan, with field becoming more vertical by $\approx 8°$ for both ROI 1 and ROI 2. After the flare, inclination decreases (*i.e.* becomes more horizontal) by $\approx 10°$ for ROI 1 and $\approx 4°$ for ROI 2. These results support the idea that the field in both ROIs becomes more vertical $\approx 6.5 - 2.5$ hours before the GOES B1.0 flare and more horizontal within $\approx$ one hour after the flare has ended. It is interesting to note that the location of the field change is near the neutral line with the plage region, in a negative-polarity region to the SE of the sunspot.

To put the changes in field parameters observed over the scans into context, it is worth estimating where the field lines in ROI 1 and ROI 2 are connected to by examining the direction of the transverse magnetic-field vectors (overlaid on the inclination scans in Figure 2). However, the true connectivity cannot be determined from 2D results and the necessary 3D extrapolations of the region will be investigated in a future article. As a first guess toward the possible connectivity, the field in ROI 1 seems to be generally pointing toward a northerly direction in Scan 1 and Scan 2, becoming increasingly more NE in Scan 3 and Scan 4. In ROI 2, the field is pointing in a general NE direction in the first scan, pointing in an increasingly more easterly direction as time progresses, finally becoming more NE after the flare. It seems that the plage region SE of the ROIs extends toward the NW (*i.e* between the ROIs) as the scans progress, before pinching off after the flare. It is difficult to determine by eye exactly where the field may be connected to over the scans, especially if relying on median values of small groups of pixels. We surmise a region of plage NE of ROI 1 to be a likely connection point. The fourth scan in Figure 2 also indicates a possible connection between ROI 2 and the portion of intersecting plage that first extends between the ROIs before "pinching off" after the flare. Studying the field distributions within the ROIs is necessary to fully understand the evolution.

3.3. Signatures of Non-potentiality

The vertical current density was calculated by the method of Crouch and Barnes (2008), as implemented in the AMBIG code. Median values of all pixels within the contours for each ROI are presented in Figure 4, with vertical bars again showing the ROI standard deviation. A familiar trend is seen between the first and second scans (*i.e.* no change within the spread of values in either ROI). Negative vertical current density increases in magnitude in the pre-flare state from the second to third scans by ≈ 0.11 mA cm^{-2} for ROI 1 and by ≈ 0.03 mA cm^{-2} for ROI 2. The magnitude subsequently decreases by ≈ 0.07 mA cm^{-2} in both ROI 1 and ROI 2. Changes in ROI 1 parameters are much more distinct than in ROI 2, as was also seen in field inclination and vertical field strength. Thus, stronger currents appear in both regions before the flare occurs, dropping back to earlier background values after the flare. An increase in current density before the flare indicates an emergence or build-up of non-potentiality in the field, with these observed changes likely to be linked to energy build-up before the flare, and energy release during to the flare.

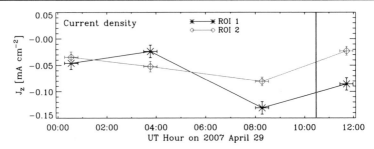

Figure 4 Temporal variation in the median values of vertical current density. Values for ROI 1 are plotted with black asterisks, and ROI 2 with blue diamonds. Vertical bars indicate the standard deviation of the property within the ROI, while horizontal bars delimit the scan duration. The vertical line between the third and fourth scans marks the flare peak time.

4. Discussion and Conclusions

An $\approx 8°$ change in field inclination toward the vertical was found leading up to the flare, with a $\approx 7°$ return toward the horizontal afterwards. Note that the inclination changes toward the vertical had occurred by ≈ 2.5 hours before the flare onset, with no changes observed $\approx 6.5 - 10$ hours beforehand. Schrijver (2007) states that the energy build-up phase can last for as much as a day in an active region, so it is interesting to see such short time-scale changes. Previous studies have also reported changes in field orientation after a flare. Li *et al.* (2009) found an inclination angle change of $\approx 5°$ toward the horizontal in a region of enhanced G-band intensity after an X-class flare, and the inclination becoming more vertical by $\approx 3°$ in a region of diminished G-band intensity. Although their study focuses on penumbral regions, the region becoming more horizontal after the flare is located close to the flaring neutral line, similar to our findings. This concept is also mentioned in some theoretical studies; *e.g.* Hudson, Fisher, and Welsch (2008) predicted that the photospheric magnetic fields close to the neutral line would become more horizontal in a simple flare-restructuring model.

Examining previous findings of transverse-field changes, Wang *et al.* (2002) used vector magnetogram observations to find an impulsive increase of the transverse-field strength and magnetic shear after three X-class flares. Li *et al.* (2009) also found a transverse-field increase of 20% after an X3.4 flare. We found no significant changes in transverse-field strength either immediately before or after the flare. However, the difference in active regions must be noted, with the Wang *et al.* and Li *et al.* works focusing on higher-magnitude flares from δ sunspot groups. Our insignificant changes in the transverse-field strength are explained by the competing field-strength and inclination changes before and after the flare. For example, a large increase in inclination angle for ROI 1 (Figure 3(d)) between the second and third scans is accompanied by a slight increase in field strength (Figure 3(a)), giving approximately no change in the horizontal field (Figure 3(b)). The Li *et al.* (2009) result supports the reconnection picture of Liu *et al.* (2005), whereby newly connected fields near the magnetic neutral line contributed to field inclination becoming more horizontal. This picture suggests that the field lines after the flare in our study become newly reconnected, low-lying, more horizontal field lines near the flaring neutral line.

Vertical-field magnitude was found in our results to increase in both ROIs before the flare, and decrease by approximately the same amount afterwards. Wang *et al.* (2002) examined LOS magnetograms as well as vector data, finding an increase in magnetic flux of the leading polarity in six X-class flares. Sudol and Harvey (2005) used longitudinal magnetogram data from the Global Oscillation Network Group to find abrupt and permanent changes in the

LOS magnetic field after 15 X-class flares. They found decreases in vertical field twice as often as increases: in 75% of cases the magnetic-field change occurred in less than ten minutes. Sudol and Harvey quote median LOS field changes of 90 G and found that the strongest field changes typically occur in penumbrae. This behavior of decreasing vertical field is reflected in our findings, although we observe larger changes of ≈ 330 G in a region outside of the penumbra.

Our observations find an increase in negative vertical current density within $\approx 6.5 - 2.5$ hours before the flare, with a decrease toward the initial pre-flare values after the flare. Strong emerging currents have often been linked with flare triggers; *e.g.* Su *et al.* (2009) observed the current density for the same active region three days later when a C8.5 flare occurred, finding strong currents along the field lines. Canou and Amari (2010) also examined the vertical current density for the same active region using a different extrapolation method from Su *et al.* (2009). The extrapolation found footpoints of the twisted flux ropes to be anchored in a region of significant vertical current (*i.e.* in the core of the flux region rather than along the field lines). They observed the breakdown of the force-free assumption along the neutral line due to non-zero vertical current density and suggested that this could be due to the emergence of the twisted flux ropes, or perhaps the presence of non-null magnetic forces. They also determined that enough free magnetic energy existed to power the C8.5 flare studied by Su *et al.* (2009) and a C4.2 flare a few days later. Similar mechanisms could possibly be at work to cause the earlier lower-magnitude flare examined in this article.

Régnier and Priest (2007) noted the discrepancies that exist between using different extrapolation methods. They found that strong currents present in the magnetic configuration were responsible for highly twisted and sheared field lines in a decaying active region. In contrast, weak currents existed in a newly emerged active region. They also suggest a strong dependance of vertical current density on the nature of the active region, *e.g.* the stage of the regions' evolution or the distribution of the sources of magnetic field. Most previous work has focused on considerably more complex active regions that produce M- or X-class flares, so it is important to note that distinct changes in the magnetic field were still observed for this B-class flare.

It is worth mentioning that Okamoto *et al.* (2009) observed converging motions in Ca II H movies of the same active region as this article, which they describe as driven by moat flows from the sunspot toward the trailing plage neutral line (*i.e.* near our ROI locations). Schrijver and Zwaan (2000) mention typical spatial scales of moat flow regions of $\approx 10-20$ Mm measured from the outer edge of the penumbra. This suggests that our ROIs lie on the outer edge of the moat flow region, and perhaps a moving magnetic feature was being driven toward the plage region. This driving would cause the field near the neutral line to become more vertical before the flare, as per our results, and might explain the pre-flare energy build-up phase. The field would then relax and become more horizontal after the energy release, as we found. The driving could be related to converging motions toward the neutral line highlighted in a number of MHD simulations (*e.g.*, Amari *et al.*, 2003). Amari *et al.* mention a three-part magnetic structure associated with their model's disruption phase, with a twisted flux rope running through a global arcade and above small loops. These newly formed small loops, described as due to reconnection, are perhaps indicative of the more horizontally inclined post-flare field of this article compared to pre-flare build-up values.

Further work is planned to clarify the connectivity of the two regions of interest and changes in the 3D topology. Our resulting disambiguated field vector will be used as an input to a magnetic-field extrapolation to determine various topology measures (*e.g.* numbers and locations of nulls, separatrix layers). However, it is interesting to see such clear changes in field-vector characteristics (such as inclination, magnetic divergence, and vertical current density) leading up to and after the flare, before making high-order calculations of 3D topology. These forms of field-orientation changes could prove to be useful precursors for flare forecasting in the future.

Acknowledgements *Hinode* is a Japanese mission developed and launched by ISAS/JAXA, with NAOJ as domestic partner and NASA and STFC (UK) as international partners. It is operated by these agencies in co-operation with ESA and NSC (Norway). S.A.M. is supported by the AXA Research Fund while D.S.B. is supported by the European Community (FP7) under a Marie Curie Intra-European Fellowship for Career Development.

References

Amari, T., Luciani, J.F., Aly, J.J., Mikic, Z., Linker, J.: 2003, *Astrophys. J.* **585**, 1073. doi:10.1086/345501.

Antiochos, S.K.: 1998, *Astrophys. J. Lett.* **502**, L181. doi:10.1086/311507.

Aschwanden, M.J.: 2005, *Physics of the Solar Corona. An Introduction with Problems and Solutions* (2nd edn.), Springer, New York.

Canfield, R.C., Pevtsov, A.A.: 1998, In: Balasubramaniam, K.S., Harvey, J., Rabin, D. (eds.) *Synoptic Solar Physics* **CS-140**, Astron. Soc. Pac., San Francisco, 131.

Canfield, R.C., Leka, K.D., Wülser, J.: 1991, In: Uchida, Y., Canfield, R.C., Watanabe, T., Hiei, E. (eds.) *Flare Physics in Solar Activity Maximum 22, Lecture Notes in Physics* **387**, Springer, Berlin, 96. doi:10.1007/BFb0032602.

Canfield, R.C., de La Beaujardiere, J., Fan, Y., Leka, K.D., McClymont, A.N., Metcalf, T.R., Mickey, D.L., Wuelser, J., Lites, B.W.: 1993, *Astrophys. J.* **411**, 362. doi:10.1086/172836.

Canou, A., Amari, T.: 2010, *Astrophys. J.* **715**, 1566. doi:10.1088/0004-637X/715/2/1566.

Chandra, R., Schmieder, B., Aulanier, G., Malherbe, J.M.: 2009, *Solar Phys.* **258**, 53. doi:10.1007/s11207-009-9392-z.

Charbonneau, P.: 1995, *Astrophys. J. Suppl.* **101**, 309. doi:10.1086/192242.

Conlon, P.A., Gallagher, P.T., McAteer, R.T.J., Ireland, J., Young, C.A., Kestener, P., Hewett, R.J., Maguire, K.: 2008, *Solar Phys.* **248**, 297. doi:10.1007/s11207-007-9074-7.

Conlon, P.A., McAteer, R.T.J., Gallagher, P.T., Fennell, L.: 2010, *Astrophys. J.* **722**, 577. doi:10.1088/0004-637X/722/1/577.

Crouch, A.D., Barnes, G.: 2008, *Solar Phys.* **247**, 25. doi:10.1007/s11207-007-9096-1.

Freeland, S.L., Handy, B.N.: 1998, *Solar Phys.* **182**, 497.

Gallagher, P.T., Moon, Y., Wang, H.: 2002, *Solar Phys.* **209**, 171. doi:10.1023/A:1020950221179.

Gary, G.A., Hagyard, M.J.: 1990, *Solar Phys.* **126**, 21.

Georgoulis, M.K., LaBonte, B.J.: 2007, *Astrophys. J.* **671**, 1034. doi:10.1086/521417.

Hewett, R.J., Gallagher, P.T., McAteer, R.T.J., Young, C.A., Ireland, J., Conlon, P.A., Maguire, K.: 2008, *Solar Phys.* **248**, 311. doi:10.1007/s11207-007-9028-0.

Hudson, H.S., Fisher, G.H., Welsch, B.T.: 2008, In: Howe, R., Komm, R.W., Balasubramaniam, K.S., Petrie, G.J.D. (eds.) *Subsurface and Atmospheric Influences on Solar Activity* **CS-383**, Astron. Soc. Pac., San Francisco, 221.

Kosugi, T., Matsuzaki, K., Sakao, T., Shimizu, T., Sone, Y., Tachikawa, S., Hashimoto, T., Minesugi, K., Ohnishi, A., Yamada, T., Tsuneta, S., Hara, H., Ichimoto, K., Suematsu, Y., Shimojo, M., Watanabe, T., Shimada, S., Davis, J.M., Hill, L.D., Owens, J.K., Title, A.M., Culhane, J.L., Harra, L.K., Doschek, G.A., Golub, L.: 2007, *Solar Phys.* **243**, 3. doi:10.1007/s11207-007-9014-6.

Lagg, A., Woch, J., Krupp, N., Solanki, S.K.: 2004, *Astron. Astrophys.* **414**, 1109. doi:10.1051/0004-6361:20031643.

Leka, K.D., Barnes, G., Crouch, A.: 2009, In: Lites, B., Cheung, M., Magara, T., Mariska, J., Reeves, K. (eds.) *The Second Hinode Science Meeting: Beyond Discovery – Toward Understanding* **CS-415**, Astron. Soc. Pac., San Francisco, 365.

Leka, K.D., Barnes, G., Crouch, A.D., Metcalf, T.R., Gary, G.A., Jing, J., Liu, Y.: 2009, *Solar Phys.* **260**, 83. doi:10.1007/s11207-009-9440-8.

Li, J., Metcalf, T.R., Canfield, R.C., Wuelser, J., Kosugi, T.: 1997, *Astrophys. J.* **482**, 490. doi:10.1086/304131.

Li, Y., Jing, J., Tan, C., Wang, H.: 2009, *Sci. China Ser. G* **52**, 1702. doi:10.1007/s11433-009-0238-3.

Liu, C., Deng, N., Liu, Y., Falconer, D., Goode, P.R., Denker, C., Wang, H.: 2005, *Astrophys. J.* **622**, 722. doi:10.1086/427868.

Liu, Y., Zhang, H.: 2001, *Astron. Astrophys.* **372**, 1019. doi:10.1051/0004-6361:20010550.

Mandrini, C.H.: 2006, In: Choudhary, D.P., Sobotka, M. (eds.) *IAU Joint Discussion 3 – Solar Active Regions and 3D Magnetic Structure, Highlights in Astronomy* **14**, Cambridge Univ. Press, New York, 3.

Metcalf, T.R.: 1994, *Solar Phys.* **155**, 235. doi:10.1007/BF00680593.

Metcalf, T.R., Canfield, R.C., Hudson, H.S., Mickey, D.L., Wulser, J., Martens, P.C.H., Tsuneta, S.: 1994, *Astrophys. J.* **428**, 860. doi:10.1086/174295.

Metcalf, T.R., Leka, K.D., Barnes, G., Lites, B.W., Georgoulis, M.K., Pevtsov, A.A., Balasubramaniam, K.S., Gary, G.A., Jing, J., Li, J., Liu, Y., Wang, H.N., Abramenko, V., Yurchyshyn, V., Moon, Y.: 2006, *Solar Phys.* **237**, 267. doi:10.1007/s11207-006-0170-x.

Okamoto, T.J., Tsuneta, S., Lites, B.W., Kubo, M., Yokoyama, T., Berger, T.E., Ichimoto, K., Katsukawa, Y., Nagata, S., Shibata, K., Shimizu, T., Shine, R.A., Suematsu, Y., Tarbell, T.D., Title, A.M.: 2009, *Astrophys. J.* **697**, 913. doi:10.1088/0004-637X/697/1/913.

Petrie, G., Sudol, J.J.: 2009, *Bull. Am. Astron. Soc.* **40**, 26.

Pevtsov, A.A., Canfield, R.C., Zirin, H.: 1996, *Astrophys. J.* **473**, 533. doi:10.1086/178164.

Priest, E.R., Forbes, T.G.: 2002, *Astron. Astrophys. Rev.* **10**, 313. doi:10.1007/s001590100013.

Régnier, S., Canfield, R.C.: 2006, *Astron. Astrophys.* **451**, 319. doi:10.1051/0004-6361:20054171.

Régnier, S., Priest, E.R.: 2007, *Astron. Astrophys.* **468**, 701. doi:10.1051/0004-6361:20077318.

Rust, D.M.: 1992, *Adv. Space Res.* **12**, 289. doi:10.1016/0273-1177(92)90119-I.

Rust, D.M., Sakurai, T., Gaizauskas, V., Hofmann, A., Martin, S.F., Priest, E.R., Wang, J.: 1994, *Solar Phys.* **153**, 1. doi:10.1007/BF00712489.

Schmieder, B., Demoulin, P., Aulanier, G., Golub, L.: 1996, *Astrophys. J.* **467**, 881. doi:10.1086/177662.

Schrijver, C.J.: 2007, *Astrophys. J. Lett.* **655**, L117. doi:10.1086/511857.

Schrijver, C.J.: 2009, *Adv. Space Res.* **43**, 739. doi:10.1016/j.asr.2008.11.004.

Schrijver, C.J., Zwaan, C.: 2000, *Solar and Stellar Magnetic Activity*, Cambridge Univ. Press, Cambridge.

Su, Y., van Ballegooijen, A., Lites, B.W., Deluca, E.E., Golub, L., Grigis, P.C., Huang, G., Ji, H.: 2009, *Astrophys. J.* **691**, 105. doi:10.1088/0004-637X/691/1/105.

Sudol, J.J., Harvey, J.W.: 2005, *Astrophys. J.* **635**, 647. doi:10.1086/497361.

Suematsu, Y., Tsuneta, S., Ichimoto, K., Shimizu, T., Otsubo, M., Katsukawa, Y., Nakagiri, M., Noguchi, M., Tamura, T., Kato, Y., Hara, H., Kubo, M., Mikami, I., Saito, H., Matsushita, T., Kawaguchi, N., Nakaoji, T., Nagae, K., Shimada, S., Takeyama, N., Yamamuro, T.: 2008, *Solar Phys.* **249**, 197. doi:10.1007/s11207-008-9129-4.

Tanaka, K.: 1986, *Astrophys. Space Sci.* **118**, 101. doi:10.1007/BF00651116.

Tsuneta, S., Ichimoto, K., Katsukawa, Y., Nagata, S., Otsubo, M., Shimizu, T., Suematsu, Y., Nakagiri, M., Noguchi, M., Tarbell, T., Title, A., Shine, R., Rosenberg, W., Hoffmann, C., Jurcevich, B., Kushner, G., Levay, M., Lites, B., Elmore, D., Matsushita, T., Kawaguchi, N., Saito, H., Mikami, I., Hill, L.D., Owens, J.K.: 2008, *Solar Phys.* **249**, 167. doi:10.1007/s11207-008-9174-z.

Unno, W.: 1956, *Publ. Astron. Soc. Japan* **8**, 108.

Wang, H., Ewell, M.W. Jr., Zirin, H., Ai, G.: 1994, *Astrophys. J.* **424**, 436. doi:10.1086/173901.

Wang, H., Spirock, T.J., Qiu, J., Ji, H., Yurchyshyn, V., Moon, Y., Denker, C., Goode, P.R.: 2002, *Astrophys. J.* **576**, 497. doi:10.1086/341735.

Zirin, H., Wang, H.: 1993, *Nature* **363**, 426. doi:10.1038/363426a0.

Solar Phys (2012) 277:59–76
DOI 10.1007/s11207-011-9907-2

Global Forces in Eruptive Solar Flares: The Lorentz Force Acting on the Solar Atmosphere and the Solar Interior

G.H. Fisher · D.J. Bercik · B.T. Welsch · H.S. Hudson

Received: 4 April 2011 / Accepted: 11 November 2011 / Published online: 9 December 2011
© Springer Science+Business Media B.V. 2011

Abstract We compute the change in the Lorentz force integrated over the outer solar atmosphere implied by observed changes in vector magnetograms that occur during large, eruptive solar flares. This force perturbation should be balanced by an equal and opposite force perturbation acting on the solar photosphere and solar interior. The resulting expression for the estimated force change in the solar interior generalizes the earlier expression presented by Hudson, Fisher, and Welsch (*Astron. Soc. Pac.* **CS-383**, 221, 2008), providing horizontal as well as vertical force components, and provides a more accurate result for the vertical component of the perturbed force. We show that magnetic eruptions should result in the magnetic field at the photosphere becoming more horizontal, and hence should result in a downward (toward the solar interior) force change acting on the photosphere and solar interior, as recently argued from an analysis of magnetogram data by Wang and Liu (*Astrophys. J. Lett.* **716**, L195, 2010). We suggest the existence of an observational relationship between the force change computed from changes in the vector magnetograms, the outward momentum carried by the ejecta from the flare, and the properties of the helioseismic disturbance driven by the downward force change. We use the impulse driven by the Lorentz-force change in the outer solar atmosphere to derive an upper limit to the mass of erupting plasma that can escape from the Sun. Finally, we compare the expected Lorentz-force change at the photosphere with simple estimates from flare-driven gasdynamic disturbances and from an estimate of the perturbed pressure from radiative backwarming of the photosphere in flaring conditions.

Solar Flare Magnetic Fields and Plasmas
Guest Editors: Y. Fan and G.H. Fisher

G.H. Fisher (✉) · D.J. Bercik · B.T. Welsch · H.S. Hudson
Space Sciences Laboratory, University of California, Berkeley, CA, USA
e-mail: fisher@ssl.berkeley.edu

D.J. Bercik
e-mail: bercik@ssl.berkeley.edu

B.T. Welsch
e-mail: welsch@ssl.berkeley.edu

H.S. Hudson
e-mail: hhudson@ssl.berkeley.edu

Keywords Active regions, magnetic · Coronal mass ejections, theory · Flares, dynamics · Flares, relation to magnetic field · Helioseismology, theory · Magnetic fields, corona

1. Introduction

Eruptive flares and coronal mass ejections (CMEs) result from global reconfigurations of the magnetic field in the solar atmosphere. Recently, signatures of this magnetic-field change have been detected in both vector and line-of-sight magnetograms, the maps of the vector and the line-of-sight component of the photospheric magnetic field, respectively. Is there a relationship between this measured-field change and properties of the eruptive phenomenon? What is the relationship between forces acting on the outer solar atmosphere and those acting on the photosphere and below, in the solar convection zone?

We will attempt to address these questions by considering the action of the Lorentz force over a large volume in the solar atmosphere that is consistent with observed changes in the photospheric magnetic field, and we will discuss how one can derive observationally testable limits on eruptive-flare or CME mass that are based on these force estimates. We will also provide more context for the recent result of Hudson, Fisher, and Welsch (2008), who present an estimate for the inward force on the solar interior driven by changes observed in magnetograms. We will also provide additional interpretation of the recent observational results of Wang and Liu (2010), who find from vector magnetogram observations that the force acting on the photosphere and interior is nearly always downward, and Petrie and Sudol (2010), who find similar results from a statistical study using line-of-sight magnetograms.

Finally, we will compare the downward impulse from changes in the Lorentz force with pressure impulses from heating by energetic-particle release during flares, and with radiative backwarming during flares, with the goal of describing the necessary future work to assess which physical mechanisms produce the largest change in force density at the photosphere, and hence which might be most effective in driving helioseismic waves (*e.g.*, Kosovichev, 2011) into the solar interior.

2. The Lorentz Force Acting on the Upper Solar Atmosphere

The Lorentz force per unit volume can be written as

$$\mathbf{f}_{\mathrm{L}} = \nabla \cdot \mathbf{T} = \frac{\partial T_{ij}}{\partial x_j}, \tag{1}$$

where the Maxwell stress tensor $[T_{ij}]$ is given by

$$T_{ij} = \frac{1}{8\pi}\left(2B_i B_j - B^2 \delta_{ij}\right), \tag{2}$$

and B_i and B_j each range over the three components of the magnetic field $[\mathbf{B}]$, and δ_{ij} is the Kronecker δ-function. Here, the divergence is expressed in Cartesian coordinates. To evaluate the total Lorentz-force $[\mathbf{F}_{\mathrm{L}}]$ acting on the atmospheric volume that surrounds a flaring active region, we integrate this force density over the volume, with the photospheric surface taken as the lower boundary of that volume, and with the upper boundary taken at some great height above the active region. The volume integral of the divergence in Equation (1) can be evaluated by using the divergence (Gauss's) theorem:

$$\mathbf{F}_{\mathrm{L}} \equiv \int_V \mathrm{d}^3 x \frac{\partial T_{ij}}{\partial x_j} = \oiint_{A_{\mathrm{tot}}} \mathrm{d}A \, T_{ij} n_j, \tag{3}$$

Figure 1 Schematic illustration of the volume in which the photospheric-to-coronal portions of a bipolar active region are embedded. It is assumed that at the outer surface, the magnetic field is negligibly small, and that the side-wall boundaries are sufficiently far away from the active region that they do not contribute to the Gauss's theorem surface integral. Note that at the photosphere, the outward surface normal vector $\hat{\mathbf{n}}$ points in the $-\hat{\mathbf{r}}$ direction. The red and blue colors represent the upward and downward vertical fluxes in the active region.

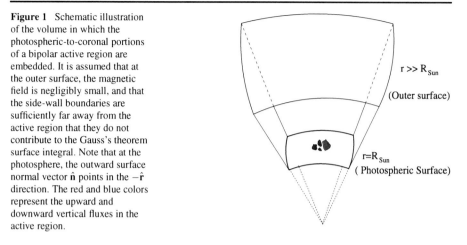

$r \gg R_{\text{Sun}}$
(Outer surface)

$r = R_{\text{Sun}}$
(Photospheric Surface)

where n_j represents the components of the outward unit vector [$\hat{\mathbf{n}}$] that is normal to the bounding surface of the atmospheric volume, and where A_{tot} represents the area of the entire bounding surface. Substituting the expression (2) for the Maxwell stress tensor results in this equation:

$$\mathbf{F}_{\text{L}} = \frac{1}{8\pi} \oiint_{A_{\text{tot}}} dA \left[2\mathbf{B}(\mathbf{B} \cdot \hat{\mathbf{n}}) - \hat{\mathbf{n}} B^2 \right]. \tag{4}$$

A sketch of the atmospheric volume surrounding the active region is shown in Figure 1.

If we assume that the upper surface of the volume is sufficiently far above the active region that the magnetic field integrated over that surface is negligible, and that the side walls are also sufficiently distant that there is negligible magnetic-field contribution from those integrals as well, then the only surface that will contribute will be the photosphere near the active region, where the magnetic fields are strong. In that case, $\hat{\mathbf{n}} = -\hat{\mathbf{r}}$ and $\mathbf{B} \cdot \hat{\mathbf{n}} = -B_r$, where B_r is the radial field component. The surface integral then results in the following two equations for the upward (*i.e.*, radial) and horizontal components of the Lorentz force [F_r] and [$\mathbf{F}_h$]:

$$F_r = \frac{1}{8\pi} \int_{A_{\text{ph}}} dA \left(B_h^2 - B_r^2 \right), \tag{5}$$

and

$$\mathbf{F}_h = -\frac{1}{4\pi} \int_{A_{\text{ph}}} dA \, B_r \mathbf{B}_h. \tag{6}$$

Here, $\mathbf{B}_h$ represents the components of $\mathbf{B}$ in the directions parallel to the photosphere, which we will henceforth refer to as the "horizontal" directions. The quantity $B_h^2 = \mathbf{B}_h \cdot \mathbf{B}_h$, and A_{ph} is the area of the photospheric domain containing the active region. If the active region is sufficiently small in spatial extent and magnetically isolated from other strong magnetic fields, one can approximate this surface integral as an integral over x and y in Cartesian coordinates, with the upward direction represented as z instead of r.

The restrictions given above regarding the side-wall contributions to the Gauss's law integrals can be relaxed if the volume of the domain is extended to a global volume: an integral over the entire outer atmosphere of the Sun. In this case, there are no side-wall boundaries to worry about, and A_{ph} coincides with the entire photospheric surface of the Sun. The outer spherical surface boundary is assumed to be sufficiently far from the Sun that it makes no significant contribution to the Gauss's law surface integral. Thus, integrals over the entire solar surface of Equations (5) and (6) should represent the total Lorentz-force acting on the Sun's outer atmosphere. If one sets the total Lorentz-force to zero, the above surface integrals yield well-known constraint equations on force-free fields (Low, 1985), with the Cartesian version of the equations being used to test the force-free condition of photospheric and chromospheric vector magnetograms (Metcalf *et al.*, 1995). If the magnetic-field distribution is not force-free, but the atmosphere is observed to be static, then presumably the Lorentz forces are balanced by other forces such as gas-pressure gradients and gravity.

Wang and Liu (2010) have found from an analysis of eleven large (X-class) flaring active regions that the vector magnetic field is always observed to change after a flare in the sense that the magnetic field becomes "more horizontal" than it was before the flare. The change is observed to occur on a time scale of a few minutes, and in some cases as fast as the sample spacing (one minute) permits. What are the implications of this observational result on the Lorentz force acting on the solar atmosphere?

To address this question, we first take the temporal derivative of Equations (5) and (6) to find

$$\frac{\partial F_r}{\partial t} = \frac{1}{8\pi} \int_{A_{\text{ph}}} dA \frac{\partial}{\partial t} (B_h^2 - B_r^2), \tag{7}$$

and

$$\frac{\partial \mathbf{F}_h}{\partial t} = -\frac{1}{4\pi} \int_{A_{\text{ph}}} dA \frac{\partial}{\partial t} (B_r \mathbf{B}_h). \tag{8}$$

Next, we assume the fields are observed to change over a time duration δt, and integrate the temporal derivatives of the Lorentz-force contributions to find the changes to the Lorentz force components δF_r and $\delta \mathbf{F}_h$:

$$\delta F_r = \frac{1}{8\pi} \int_{A_{\text{ph}}} dA (\delta B_h^2 - \delta B_r^2), \tag{9}$$

and

$$\delta \mathbf{F}_h = -\frac{1}{4\pi} \int_{A_{\text{ph}}} dA \delta (B_r \mathbf{B}_h), \tag{10}$$

where, at a fixed location in the photosphere,

$$\delta B_h^2 \equiv \int_0^{\delta t} dt \frac{\partial}{\partial t} B_h^2 = B_h^2(\delta t) - B_h^2(0), \tag{11}$$

$$\delta B_r^2 \equiv \int_0^{\delta t} dt \frac{\partial}{\partial t} B_r^2 = B_r^2(\delta t) - B_r^2(0), \tag{12}$$

$$\delta (B_r \mathbf{B}_h) \equiv \int_0^{\delta t} dt \frac{\partial}{\partial t} (B_r \mathbf{B}_h) = B_r(\delta t) \mathbf{B}_h(\delta t) - B_r(0) \mathbf{B}_h(0). \tag{13}$$

These quantities are simply the observed changes in the magnetic variables that occur over the course of a flare. Note that if the flaring active region is near disk center, so that the observed transverse magnetic field is a good approximation to the horizontal field, then the expression for δF_r can be evaluated without having to perform the 180° disambiguation of the vector magnetogram data – only the amplitude of the horizontal field enters into the expression.

62

⌃ Springer

If the change in the Lorentz force is significant only within small areas of the photo-sphere, then the only contribution to the global-area integral will be from the smaller do-mains where the changes are significant, potentially simplifying the evaluation of Equa-tions (9) and (10).

We now argue that in the outer atmosphere of flaring active regions, the impulse from the changed Lorentz force dominates all other forces. First, from energetic considerations, the magnetic field is believed to be the source of energy for eruptive flares and CMEs: Forbes (2000) and Hudson (2007) have argued that no other known source of energy can provide the observed kinetic energy of outward motion observed in CMEs, and there simply is no other viable source for the thermal and radiated energy known to be released in solar flares. Second, apart from the Lorentz force, the only other significant forces known to operate on the solar atmosphere are gas-pressure gradients and gravity. To evaluate the change in the gas-pressure gradient forces in the outer atmosphere, one can perform a Gaussian volume integral over the outer solar atmosphere of the vertical component of the gas-pressure gradi-ent force. The net change in the vertical force is just the difference between the gas-pressure change at the top of the Gaussian volume from that at the bottom. If the plasma β in the solar atmosphere is low, as is generally the case in active regions, it seems unlikely that this will be as significant as the change of the Lorentz force. Nevertheless, in Section 3, we will consider perturbations to the gas pressure at the photosphere and discuss their effectiveness. In the case of the gravitational force, unless the plasma has moved a huge distance ($\approx R_\odot$) away from the Sun on the time scale of the observed field change, the gravitational force acting on the given mass of the plasma within the Gaussian volume must be approximately the same, and hence the change in the gravitational force should be small.

The results of Wang and Liu (2010), in which the field becomes more horizontal after the occurrence of eruptive flares, are thus consistent with an upward impulse acting on the outer atmosphere, so we identify this impulse as the photospheric magnetic-field signature of the force driving a magnetic eruption. To estimate the magnitude of the impulse, we make the simple assumption that the change in the Lorentz force in Equations (9) and (10) occurs linearly with time from $t = 0$ to $t = \delta t$. We denote the mass of the plasma that is eventually ejected as M_{ejecta}, and we assume the fluid velocity averaged over this plasma to be zero prior to the eruption. The Lorentz impulse will then be related to the ejecta's momentum by

$$\frac{1}{2}\delta F_r \delta t = M_{\text{ejecta}} v_r \tag{14}$$

and

$$\frac{1}{2}\delta \mathbf{F}_h \delta t = M_{\text{ejecta}} \mathbf{v}_h, \tag{15}$$

where v_r is the upward (radial) component of the velocity of the ejecta after the impulse, and $\mathbf{v}_h$ is the resulting horizontal component of the ejecta velocity. Note that if v_r is less than the escape velocity [$v_e \equiv (2GM_\odot/R_\odot)^{1/2}$], the ejecta will ultimately be stopped by gravity and will not result in an eruption. If v_r exceeds v_e, we assume that the ejecta can become a coronal mass ejection (CME). This means that for a given observation of magnetic-field changes in a flaring active region, there is an upper limit to the mass of any resulting CME given by

$$M_{\text{CME}} < \frac{1}{2}\frac{\delta F_r \delta t}{v_e}. \tag{16}$$

For the magnetic-field changes in the 2 November 2003 flare studied by Hudson, Fisher, and Welsch (2008), they estimate a change in the Lorentz-force surface density of

2500 dyne cm^{-2}, and with the estimated area over which the change occurs, a total force of $\approx 10^{22}$ dyne. This is probably an underestimate for the total upward force, since this case was taken from the study by Sudol and Harvey (2005), in which only the line-of-sight contributions were measured. A more extensive analysis of a larger data set of flares (Petrie and Sudol, 2010) has since shown several other cases of comparable or even larger Lorentz forces for some X-class flares. Assuming a time scale of ten minutes for the photospheric magnetic fields to change (from the temporal evolution results of Sudol and Harvey, 2005 and Petrie and Sudol, 2010) then results in an upper limit on the mass of any CME coming from this flaring active region of 4.9×10^{16} g.

An expression for the Lorentz-driven impulse in the horizontal directions (Equation (15)) could be useful in determining the initial deflection of the ejecta away from a radial trajectory. The initial trajectory direction can be determined by examining the ratio of the horizontal components of $\delta \mathbf{F}_h$ to δF_r.

What effect do these Lorentz-force changes, and their resulting impulses, have on the response of the solar-interior plasma at and below the photosphere? To estimate the change in the Lorentz force acting on the solar interior (the interior is defined here to be the plasma that extends from the photosphere downward), one can perform almost exactly the same Gaussian volume exercise as above, but using a subsurface volume instead of an outer atmosphere volume. By performing the global integral of the Lorentz-force density over the entire volume below the photosphere, one can see that both the absolute Lorentz forces (Equations (5) and (6)) and the changes in the Lorentz forces (Equations (9) and (10)) involve exactly the same photospheric surface terms as for the outer solar atmosphere, except that the outward surface normal $\hat{\mathbf{n}}$ is in the $+\hat{\mathbf{r}}$ direction instead of in the $-\hat{\mathbf{r}}$ direction. Thus the three components of the Lorentz force, and the flare-induced changes in the Lorentz force, have exactly the same magnitude, but opposite sign from the Lorentz forces acting on the solar atmosphere – the Lorentz-force changes acting on the interior and the solar atmosphere are exactly balanced:

$$\delta F_{\text{r,interior}} = \frac{1}{8\pi} \int_{A_{\text{ph}}} \mathrm{d}A \left(\delta B_r^2 - \delta B_h^2 \right), \tag{17}$$

and

$$\delta \mathbf{F}_{\text{h,interior}} = \frac{1}{4\pi} \int_{A_{\text{ph}}} \mathrm{d}A \delta (B_r \mathbf{B}_h). \tag{18}$$

The radial component of the force change acting on the solar interior was identified as a magnetic jerk by Hudson, Fisher, and Welsch (2008).

To relate these results (Equations (17)–(18)) to those of Hudson, Fisher, and Welsch (2008), we note that if we let z be in the upward direction, and use x and y to denote the horizontal directions, and further make the first-order approximation that $\delta B_h^2 \approx 2B_x \delta B_x + 2B_y \delta B_y$, and that $\delta B_z^2 \approx 2B_z \delta B_z$, where δB_x, δB_y, and δB_z are the observed changes in B_x, B_y, and B_z, then Equation (17) yields the unnumbered expression given by Hudson, Fisher, and Welsch (2008), assumed to be integrated over the vector magnetogram area:

$$\delta F_{\text{z,interior}} \approx \frac{1}{4\pi} \int \mathrm{d}A (-B_x \delta B_x - B_y \delta B_y + B_z \delta B_z). \tag{19}$$

If Equation (19) is evaluated over the flaring active region, such that surface terms on the vertical side walls make no significant contributions to the Gaussian integral, and the amplitude of the field-component changes is small compared to their initial values, then this expression should be robust and accurate. For future investigations of vector-magnetogram data, we believe that Equations (17)–(18) will generally be more useful than Equation (19) since they do not assume the first-order approximation, and the horizontal components of the force are included.

Since we assert that the Lorentz-force change is the dominant force acting in the outer solar atmosphere, and that this force drives an eruptive impulse, it follows from conservation of momentum that an equal and opposite impulse must be applied on the plasma in the solar interior and that, at least initially, the force driving this impulse is the Lorentz force identified in Equations (17) and (18). However, we expect that once the impulse has penetrated more than a few pressure scale heights into the solar interior, the disturbance will propagate mainly as a gasdynamic-pressure disturbance (acoustic wave), since the plasma β is thought to increase very rapidly with depth below the photosphere. For a more general discussion about momentum balance issues in solar flares, see Hudson *et al.* (2011), included in this topical issue.

Putting all of this together, we suggest that there is an observationally testable relationship between the measured Lorentz-force change and the outward momentum of the erupting ejecta that occurs over the course of an eruptive flare, and that the Lorentz force responsible for the eruption should also drive a downward-moving impulse into the solar interior. The downward-moving impulse could potentially be the source of observed "sunquake" acoustic emission detected with helioseismic techniques (Kosovichev and Zharkova, 1998; Moradi *et al.*, 2007; Kosovichev, 2011) for some solar flares. Thus we suggest the possibility of using helioseismology to study eruptive solar flares, if the detailed wave mechanics of the impulse moving downward into the interior can be better characterized and understood.

3. Other Disturbances in the Force

We argue above that assuming that the plasma β in the flaring active region is small implies that changes to gas-pressure gradients during a flare are probably unimportant compared to changes in the Lorentz force. Nevertheless, the flare-induced gas-pressure change from energy deposited in the flare atmosphere has been considered to be a viable candidate for the agent that excites flare-associated helioseismic disturbances (Kosovichev and Zharkova, 1995). Another suggested mechanism is heating near the solar photosphere driven by radiative backwarming of strong flaring emission occurring higher up in the solar atmosphere (Donea *et al.*, 2006; Moradi *et al.*, 2007; Lindsey and Donea, 2008). We consider each of these possibilities in the following sections.

3.1. Pressure Changes Driven by Flare Gasdynamic Processes

During the impulsive phase of flares, emission in the hard X-ray and γ-ray energy range is typically emitted from small, rapidly moving kernels in the chromosphere of the flaring active region (*e.g.*, Fletcher *et al.*, 2007). This emission is generally assumed to be the signature of energy release in the form of a large flux of energetic electrons. Energetic electrons in the $10-100$ KeV range that impinge on the solar atmosphere will emit nonthermal bremsstrahlung radiation from Coulomb collisions with the ambient ions in the atmosphere, and will also rapidly lose energy via Coulomb collisions with ambient electrons, resulting in strong atmospheric heating (Brown, 1971). This results, in turn, in a large gas-pressure increase in the upper chromosphere, due to rapid chromospheric evaporation. Kosovichev and Zharkova (1995) proposed that this large pressure increase is responsible for the flare-driven helioseismic waves into the solar interior that have been observed.

Can this pressure increase in the flare chromosphere result in a sufficiently great pressure change at the photosphere to be significant compared to the observed changes in the

Lorentz force? To investigate this question, we show that any gasdynamic disturbance that reaches the photosphere will propagate first as a shock-like disturbance (a "chromospheric condensation") followed by the propagation of a weaker disturbance that can be treated in the acoustic limit ($v/c_s \ll 1$). We then describe the work necessary to determine whether this acoustic disturbance has a perturbed pressure that can be comparable in strength to the Lorentz-force disturbance that we considered in Section 2. The first task is to estimate the pressure increase in the flare chromosphere that drives the chromospheric condensation.

The pressure increase in the flare chromosphere occurs when plasma that was originally at chromospheric densities is heated to coronal temperatures. The size of the pressure increase depends on the details of how the heating is applied to the pre-flare atmosphere. If the flux of nonthermal electrons is increased very suddenly, and with a sufficiently great amplitude to exceed the maximum ability of the upper chromosphere to radiate away the nonthermal electron energy flux, the result is "explosive evaporation" (Fisher, Canfield, and McClymont, 1985b, 1985c). In this case, the location of the flare transition region moves very quickly to a significantly greater depth in the atmosphere. The column depth of this location can be determined by applying the suggestion of Lin and Hudson (1976), equating the nonthermal-electron heating rate with the maximum radiative cooling rate, assuming a transition-region temperature. The validity of this approach was subsequently verified in the numerical simulations of Fisher, Canfield, and McClymont (1985c). The plasma between the original and flare transition-region column depths then explodes, driving violent mass motion both upward and downward.

Fisher (1987) developed an analytical model for the explosive evaporation process, including estimates for the maximum pressure achieved during explosive evaporation, in terms of the portion of the total nonthermal electron energy flux [F_{evap}] that is deposited between the original and flare transition-region column depths. In Figure 2, we explore explosive evaporation by first showing the computed ratio of F_{evap} to the total flux of nonthermal electrons F_{nte} for many possible cases of explosive evaporation, assuming a range of pre-flare atmospheric coronal pressure, values of the assumed low-energy cutoff [E_c], the electron spectral index [δ] (see Section III of Fisher, Canfield, and McClymont (1985c) for definitions of E_c and δ), and the total flux of nonthermal electrons [F_{nte}]. Pre-flare coronal pressures include a low value of 0.3 dyne cm^{-2} (diamonds), corresponding to a tenuous pre-flare coronal density, and a higher value of 3.0 dyne cm^{-2} (triangles), corresponding to a denser pre-flare corona. Values of E_c include 10 KeV (blue), 20 KeV (green), and 25 KeV (red). Electron spectral index [δ] values include 4 (solid curves), 5 (dotted curves), and 6 (dashed curves). Rather than assuming a sharp low-energy cutoff to the spectrum, which produces an unphysical cusp in the nonthermal-electron heating rate as a function of column depth, we adopt the modified form of the heating rate suggested in Figure 1 and Equations (9) – (11) of Fisher, Canfield, and McClymont (1985c), in which the heating rate varies smoothly with column depth. This implies that the low-energy cutoff [E_c] corresponds more to a spectral rollover than a true cutoff; the detailed electron spectra corresponding to this particular rollover behavior are given in Equations (46) – (50) of Tamres, Canfield, and McClymont (1986). Once the values for F_{evap} have been obtained, one can find the average per-particle heating rate in the explosively evaporating region and use Equations (38) – (39) from Fisher (1987) to compute the maximum pressure due to explosive evaporation. The resulting values are shown as the colored triangles and diamonds in Figure 3 as functions of the energy flux driving explosive evaporation.

We also compute the maximum pressure using an entirely different assumption for how chromospheric evaporation occurs. If the energy flux of nonthermal electrons increases more slowly than the time scale for which explosive evaporation occurs, or if the nonthermal electron energy is simply dumped into the corona and transition region through bulk heating,

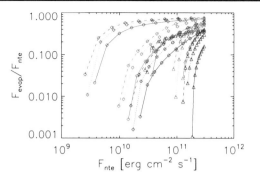

Figure 2 Fraction of the total nonthermal electron energy flux [F_{nte}] that goes into driving explosive chromospheric evaporation as a function of F_{nte}. Diamonds indicate a pre-flare coronal pressure of 0.3 dyne cm^{-2}, while triangles indicate a higher pre-flare coronal pressure of 3.0 dyne cm^{-2}. Blue symbols indicate a 10 KeV low-energy cutoff, green indicates a 20 KeV low-energy cutoff, and red indicates a 25 KeV low-energy cutoff. Solid curves indicate an electron spectral index $\delta = 4$, dotted curves $\delta = 5$, and dashed curves $\delta = 6$.

then chromospheric evaporation will still occur, and the pressure will still increase, but not as violently as assumed in the model of Fisher (1987). In the Appendix of Fisher (1989), it was shown by considering the dynamics of chromospheric evaporation driven via thermal conduction that the maximum pressure is given approximately by Equation (32) of Fisher (1989). We use this expression to compute a second estimate for the maximum pressure driven by chromospheric evaporation, using the energy flux deposited above the flare transition region from the assumed atmospheric and electron spectral characteristics described earlier. The results are shown as the black diamonds and triangles in Figure 3.

Note that both sets of estimates for the maximum pressure result in the scaling $P_{max} \sim F_{evap}^{2/3}\rho_{co}^{1/3}$, where ρ_{co} is the pre-flare coronal mass density. This result is consistent with what one might find from a simple dimensional analysis; the only difference between the two estimates is simply a different constant of proportionality that results from the detailed assumptions in the two different evaporation models.

In addition to the preceding estimates, we also plot in Figure 3 the maximum pressure as a function of the estimated energy flux driving evaporation achieved in the two largest flux simulations of Fisher, Canfield, and McClymont (1985c), two cases from Abbett and Hawley (1999), and the highest flux case from Allred *et al.* (2005). Note that, in all cases, the results from the gasdynamic simulations are either close to the approximated maximum pressures from the estimates derived above, or else are bracketed by our estimates. This is even true when the assumptions used in deriving the approximate results are not strictly adhered to in the simulations. Thus we can feel some confidence that our simpler estimates will probably bracket most cases. In particular, the explosive evaporation estimates (colored triangles and diamonds) seem to provide good upper limits to the maximum pressure due to chromospheric evaporation found from any of the simulations. By examining Figures 2 and 3, we conclude that, for large flare nonthermal electron energy fluxes $\approx 10^{11}$ erg s^{-1} cm^{-2}, the maximum pressure increase in the chromosphere is ≈ 2000 dyne cm^{-2}, and in most cases, considerably less. In summary, to achieve gas-pressure increases this high requires the highest nonthermal energy fluxes and a very rapid onset of these high-energy fluxes.

The pressure increase drives not only the rapid upward motion of the evaporating plasma into the corona, but also slower, denser flows of plasma downward into the chromosphere (*e.g.*, Fisher, Canfield, and McClymont, 1985a) described as "chromospheric condensations." These dense, downward-moving plugs of plasma form behind a downward-moving

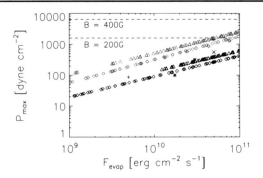

Figure 3 Maximum gas pressure driven by impulsive-phase flare heating from nonthermal electrons as a function of the energy flux driving chromospheric evaporation. The colored diamonds show the maximum pressure computed using the analytical explosive evaporation model of Fisher (1987), for a pre-flare coronal pressure of 0.3 dyne cm^{-2}, while the triangles show the maximum pressure for a pre-flare coronal pressure of 3.0 dyne cm^{-2}. Blue symbols indicate a 10 KeV low-energy cutoff, green indicates a 20 KeV low-energy cutoff, and red indicates a 25 KeV low-energy cutoff. The black symbols indicate the maximum pressure using the alternative model in which the nonthermal energy drives evaporation indirectly via thermal conduction, from Equation (32) of Fisher (1989). The black diamonds and triangles indicate the same pre-flare coronal pressures as above. Both approximations show that $P_{\max} \sim F_{\mathrm{evap}}^{2/3} \rho_{\mathrm{co}}^{1/3}$, where ρ_{co} is the pre-flare coronal mass density, but with different proportionality constants. Also plotted are the maximum pressures from several radiation-hydrodynamic flare simulations: the two highest flux cases in Fisher, Canfield, and McClymont (1985a 1985b, 1985c) (*), two cases from Abbett and Hawley (1999) (+), and a case from Allred *et al.* (2005) (×). The two dashed horizontal lines denote the magnetic pressure for field strengths of 200 G and 400 G, respectively.

shock-like disturbance driven by the pressure increase from chromospheric evaporation in flares. Simple, analytic models of the dynamic evolution of chromospheric condensations were developed by Fisher (1989). The models did a good job of describing the results of more detailed numerical gasdynamic simulations. One interesting property of the models is that, during the time period for which the downflow evolution is well described in terms of chromospheric condensations, the dynamical evolution is insensitive to the details of cooling behind the downward-moving front of the chromospheric condensation. Further, several simulation results indicate that the gas pressure in the chromospheric condensation just behind the front of the condensation is relatively constant in time, as the condensation propagates deeper into the atmosphere. This result was used in the analytical models of the condensation dynamics (Fisher, 1989). Radiative cooling immediately behind the downward-moving shock at the head of the chromospheric condensation leads to densities in the condensation that are much greater than the density ahead of it. Fisher (1989) showed by applying mass- and momentum-conservation jump conditions, plus differing assumptions about how the plasma is cooled, that the velocity evolution is very insensitive to the details of how the plasma is cooled, provided that the resulting density jump is large (*e.g.*, see the comparisons in Figure 1 of that article). These models predict the maximum column depth that the chromospheric condensation can penetrate into the solar atmosphere as a shock-like disturbance, in terms of the flare-induced pressure [$P_{\max}$] driven by electron-beam heating of the solar atmosphere. The maximum column depth of propagation [$N_{\max}$] is given approximately by

$$N_{\max} = \frac{P_{\max}}{\bar{m} g}, \qquad (20)$$

℣ Springer

where $\bar{m} \approx 1.4 m_p$ is the mean mass per proton in the solar atmosphere, and $g = 2.74 \times 10^4$ cm s^{-2} is the value of surface gravity. The values of P_{max} mentioned above result in values of N_{max} that are no larger than $\approx 3 \times 10^{22}$ cm^{-2}. However, the column depth of the solar photosphere is $\approx 10^{24}$ cm^{-2}. Thus, using the chromospheric-condensation model, flare-driven pressure disturbances can propagate to at most 3% of the column depth of the photosphere as chromospheric condensations; but this does not mean that the downflows cease at this depth: it means only that the equation of motion for chromospheric condensation (Equation (10) from Fisher (1989)) no longer applies when the driving pressure approaches the ambient pressure ahead of the condensation. At the depths where this occurs, downflow velocities become significantly less than the sound speed, and are therefore better treated in the acoustic limit.

Because the chromospheric-condensation model's assumptions begin to break down at the last stages of its evolution, we then consider the subsequent downward propagation of flare-driven pressure disturbances between column depths of $\approx 3 \times 10^{22}$ cm^{-2} and $\approx 10^{24}$ cm^{-2} using an entirely different approach: We assume that the disturbance can be represented by an acoustic wave, driven by a simple downward pulse corresponding to the last stages of the chromospheric condensation evolution. For simplicity, we assume simple, adiabatic wave evolution in an isothermal, gravitationally stratified approximation of the lower chromosphere, assuming an ideal gas equation of state, without dissipation. As described in more detail below, there are reasons to question these assumptions, but this solution allows us to demonstrate some general properties of the resulting wave evolution.

Assuming that the pre-flare chromosphere can be represented by an isothermal, gravitationally stratified atmosphere at temperature T_{ch} with pressure scale height $\Lambda_P = c_s^2/(\gamma g)$, where c_s is the adiabatic sound speed and γ is the ratio of specific heats, the equation for the perturbed vertical velocity is

$$\frac{\partial^2 v}{\partial t^2} - c_s^2 \left(\frac{1}{\Lambda_P} \frac{\partial v}{\partial s} + \frac{\partial^2 v}{\partial s^2} \right) = 0. \tag{21}$$

Here, s measures vertical distance in the downward direction, measured from the final position of the chromospheric condensation. We assume that at the depth where the chromospheric-condensation solution breaks down, the result of its final propagation is a downward displacement $[\Delta s]$ occurring over a short time. We then want to follow this displacement, using the above acoustic wave equation, as it propagates downward. At $s = 0$, we therefore assume the temporal evolution of the velocity $[v]$ to be given by

$$v(s = 0, t) = \Delta s \delta(t), \tag{22}$$

where $\delta(t)$ is the Dirac δ-function. By performing a Laplace transform of Equation (21) with this assumed time behavior at $s = 0$, we find

$$v(s, t) = \Delta s \exp\left(-\frac{s}{2\Lambda_P} \right)$$

$$\times \left[\delta\left(t - \frac{s}{c_s} \right) - \frac{s}{2\Lambda_P \sqrt{t^2 - \frac{s^2}{c_s^2}}} J_1\left(\omega_a \sqrt{t^2 - \frac{s^2}{c_s^2}} \right) H\left(t - \frac{s}{c_s} \right) \right], \tag{23}$$

where J_1 is a Bessel function, H is the Heaviside function, and ω_a is the acoustic cutoff frequency $[\omega_a = c_s/(2\Lambda_P)]$.

Note that the solution corresponds to the downward propagation of the pulse, along with a trailing wake that oscillates at a frequency that asymptotically approaches the acoustic cutoff frequency. While it is clear that the velocity amplitude decreases rapidly (there is an overall envelope function of $\exp(-s/(2\Lambda_P))$ as the pulse propagates deeper), one can show that the perturbed-pressure amplitude associated with the pulse actually increases as $\exp(s/(2\Lambda_P))$ as the pulse propagates deeper. It is therefore possible that the pressure amplitude of the acoustic wave could be even larger than the initial pressure $[P_{max}]$ of the flare chromosphere, derived above.

On the other hand, there are a number of dissipation mechanisms that our wave solution does not include, which could dramatically reduce the perturbed pressure of the resulting acoustic wave. To the extent that energy balance between radiative cooling and flare heating by the most energetic electrons is important at these depths, Fisher, Canfield, and McClymont (1985a) showed in Section IV of their article that high-frequency acoustic waves were very strongly damped by radiative cooling; low-frequency acoustic waves were damped more weakly, but still attenuated on length scales of ≈ 200 km.

To summarize, we have shown how one can estimate the peak gasdynamic pressure driven by chromospheric evaporation, and that the largest possible values of this pressure require both high-energy fluxes and rapid onset of those fluxes to produce explosive evaporation. We have also shown that the ensuing shock-like disturbance (a chromospheric condensation) can only propagate down to roughly 3% of the column depth of the photosphere, but that the dying chromospheric condensation can continue propagating downward as an acoustic wave to photospheric depths. We found a wave solution that is dispersive, consisting of both a pulse and a trailing wake, and we used a simplified example to show that the perturbed pressure of the acoustic wave can increase as the wave propagates down toward the photosphere. We then discussed a number of wave-dissipation mechanisms that may efficiently extract energy from the wave. The extent to which the gasdynamic-excited acoustic wave at the photosphere is important relative to the Lorentz-force perturbation (Section 2) will depend critically on a detailed evaluation of these wave-dissipation effects on the acoustic solution; this is beyond the scope of what we can present here.

3.2. Pressure Changes Driven by Radiative Backwarming

Observations of the spatial and temporal variation of optical continuum (white-light) emission and hard X-ray emission during solar flares show an intimate temporal, spatial, and energetic relationship between the presence of energetic electrons in the flare chromosphere and white-light emission from the solar photosphere (Hudson *et al.*, 1992; Metcalf *et al.*, 2003; Chen and Ding, 2005; Watanabe *et al.*, 2010). One possible component mechanism of this connection is radiative backwarming of the continuum-emitting layers by UV and EUV line and free–bound emission that is excited by energetic electrons penetrating into the flare chromosphere, at some distance above the photosphere. Since the radiative cooling time of the flare chromosphere immediately below the regions undergoing chromospheric evaporation is so short (Fisher, Canfield, and McClymont, 1985a), the temporal variation of UV and EUV line emission from plasma within the $10^4 - 10^5$ K temperature range will closely track heating by energetic electrons, detected as hard X-ray emission emitted from footpoints in the flare chromosphere. The backwarming scenario is illustrated schematically in Figure 4.

Estimates of the continuum opacity and atmospheric density near the solar photosphere indicate that the layer responsible for most of the optical continuum emission is about one continuum-photon mean-free path thick, or roughly 70 km. Thus most of the energy from the impinging backwarming radiation will be reprocessed into optical continuum emission within a thin layer near the solar photosphere.

Does the absorption of this radiation within this thin layer result in a significant downward force, via a pressure perturbation from enhanced heating? This mechanism has been

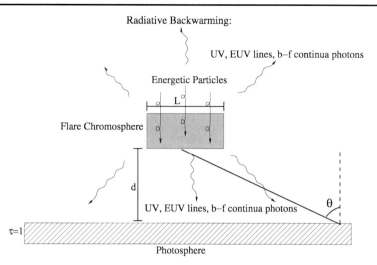

Figure 4 Schematic illustration of radiative backwarming in a solar flare. Energetic electrons are stopped collisionally in the upper flare chromosphere, raising the temperature of the plasma there. The increased energy input is balanced by an increased radiative output in the form of EUV and UV line radiation, and the emission of free–bound continua from H and other ions. This radiation is emitted in all directions, but a significant fraction of it is reabsorbed in optically thick layers near the solar photosphere. These layers respond with an increase of temperature and pressure, with an amplitude that will depend sensitively on the energy flux, area coverage, and timing of the impinging radiation. The emitting layer is assumed to be at a height d above the photosphere. The shape of the emitting layer seen from directly below is assumed to be circular, with diameter L. At an arbitrary location on the photospheric surface, the angle between the direction to the center of the source and the vertical direction is θ, and the corresponding direction cosine $\mu = \cos(\theta)$.

suggested by Donea *et al.* (2006), Moradi *et al.* (2007), and Lindsey and Donea (2008) as a potential source for "sunquake" acoustic emission seen during a few solar flares. Here, we compare and contrast this mechanism of creating a force perturbation with that from the Lorentz-force change described earlier.

The simplest estimate of the pressure change is to assume that the backheated photospheric plasma is frozen in place during the heating process, and that its temperature will rise to a level where the black-body radiated energy flux equals the combined output of the pre-flare solar radiative flux plus the incoming flare energy flux due to backwarming.

What is the flux of energy from backwarming available to heat the photosphere? In order to compute this flux, we must first estimate the fraction of the nonthermal electron energy flux that is balanced by chromospheric UV/EUV line and free–bound continuum emission, and the fraction of this radiated energy that impinges on the nearby solar photosphere. We can use the estimate presented in Section 3.1 and plotted in Figure 2 for $F_{\mathrm{evap}}/F_{\mathrm{nte}}$ to find the flux of energy [F_{rad}] that is converted from nonthermal electrons into radiated energy,

$$F_{\mathrm{rad}} \approx (1 - F_{\mathrm{evap}}/F_{\mathrm{nte}})\,F_{\mathrm{nte}}. \qquad (24)$$

Not all of this energy flux will be available for backwarming, because the resulting radiation is emitted isotropically, while the photosphere lies beneath the radiating source. If the lateral dimension L of the illuminating region is much larger than the distance d above the photosphere, then up to half the radiated energy flux will irradiate the photosphere directly beneath the source (see Figure 4). On the other hand, if the ratio d/L is of order unity, there is a substantially reduced geometrical dilution factor [f_{geom}] that must multiply F_{rad} to de-

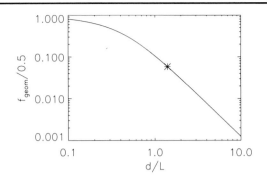

Figure 5 Computed ratio of geometrical dilution factor to the plane-parallel value of $1/2$ as a function of d/L, where d is the height of the source above the photosphere, and L is the diameter of the source, which is assumed to have a circular shape. The asterisk on the curve shows the value for $d/L = 1.4$ (see text). The plot assumes $\mu = 1$, *i.e.*, for a point on the photosphere directly beneath the irradiating source.

termine the flux of radiated energy that is incident on the photosphere beneath the source. We estimate the geometrical dilution factor [f_{geom}] as

$$f_{\text{geom}} = \frac{1}{2}\left(1 - \frac{2d/L}{\sqrt{1 + 4(d/L)^2}}\right)\mu^3, \tag{25}$$

where for simplicity this expression assumes that the shape of the irradiating source shown in Figure 4 is a circular disk of diameter L. If the position of interest on the photosphere is not directly beneath the illuminating source, but instead is located at an angle θ away from the vertical direction (see Figure 4), this expression includes a factor of μ^3, where $\mu = \cos(\theta)$, accounting for both increased distance from the source to the given point on the photosphere and the oblique angle of the irradiating source relative to the normal direction. With the geometrical dilution factor determined, this results in the following estimate of the elevated photospheric temperature [T]:

$$\sigma T^4 = \sigma T_0^4 + f_{\text{geom}} F_{\text{rad}}, \tag{26}$$

where T_0 is the non-flare photospheric effective temperature. This expression can be rewritten as

$$\frac{\Delta T}{T_0} = \left(1 + \frac{f_{\text{geom}} F_{\text{rad}}}{\sigma T_0^4}\right)^{1/4} - 1, \tag{27}$$

where $\Delta T/T_0$ is the ratio of the temperature rise to the pre-flare photospheric temperature.

Next, we estimate the height above the photosphere for the source of the backwarming radiation. To do this we find the change in depth between the pre-flare and flare transition region, using the same explosive evaporation model described in Section 3.1. For cases where the assumed flux in nonthermal electrons exceeds 10^{11} erg cm^{-2} s^{-1}, the depth of the flare transition region relative to the pre-flare transition region moves downward by distances ranging from ≈ 100 km for the dense pre-flare corona, up to ≈ 600 km for the tenuous pre-flare corona. The primary source of the backwarming radiation will be in the layers immediately below the flare transition region. Assuming an approximate distance between the photosphere and pre-flare transition region of 2000 km (*e.g.*, model F of Vernazza, Avrett, and Loeser, 1981) thus leads to an expected distance that ranges from 1400 to 1900 km between the backwarming source and the photosphere. We then estimate the geometrical dilution of a UV/EUV emitting flare kernel with roughly $L \approx 1000$ km in horizontal extent (based on

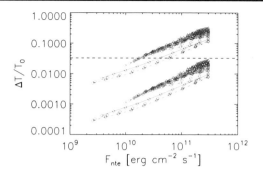

Figure 6 Computed ratio of temperature rise to background photospheric temperature as a function of the flux of energy in nonthermal electrons, using Equation (27). Diamonds indicate a pre-flare coronal pressure of $0.3\ \mathrm{dyne\,cm^{-2}}$, while triangles indicate a higher pre-flare coronal pressure of $3.0\ \mathrm{dyne\,cm^{-2}}$. Blue symbols indicate a 10 KeV low-energy cutoff, green indicates a 20 KeV low-energy cutoff, and red indicates a 25 KeV low-energy cutoff. Solid curves indicate an electron spectral index $\delta = 4$, dotted curves $\delta = 5$, and dashed curves $\delta = 6$. The geometrical dilution factor in the upper set of curves was set to $1/2$, consistent with plane-parallel geometry, in which the horizontal dimension L of the illuminating region is much greater than the height d of the illuminating region above the photosphere. The lower set of curves was computed by setting d/L to 1.4, consistent with a 1000 km diameter source illuminating the photosphere from a height of 1400 km (see text). The dashed horizontal line indicates the temperature ratio needed to match a vertical Lorentz-force surface density of $2500\ \mathrm{dyne\,cm^{-2}}$.

estimated flare-kernel areas of roughly one arcsecond^2) located in the flare chromosphere at a distance $d = 1400$ km above the photosphere, in keeping with the above distance estimates, and find a geometrical dilution factor, assuming $d/L = 1.4$, of $f_{\mathrm{geom}} = 0.029$, shown as the asterisk in Figure 5. The low value of f_{geom} results from the fact that the source as seen from the photosphere subtends a solid angle of less than 0.4 steradian, compared with the 4π steradians over which the radiation is emitted.

In Figure 6, we use Equation (27) to plot the temperature enhancement as a function of nonthermal electron-energy flux for two different assumed values of f_{geom}: $1/2$, corresponding to a widespread (plane-parallel) source of backwarming radiation, and 0.029, corresponding to $d/L = 1.4$. This figure shows a wide range of possible values of $\Delta T/T$ for a commonly assumed range of nonthermal electron energy fluxes. This fact, coupled with the wide range of possible values for f_{geom}, illustrates the difficulty in making broad conclusions about the effectiveness of backwarming in perturbing the photosphere.

The horizontal line in Figure 6 corresponds to the temperature ratio that would lead to a pressure increase comparable to the estimated Lorentz-force surface density of $2500\ \mathrm{dyne\,cm^{-2}}$ for the large flare discussed in Hudson, Fisher, and Welsch (2008). Here, we adopt a pre-flare photospheric pressure of $7.6 \times 10^4\ \mathrm{dyne\,cm^{-2}}$ (see model S of Christensen-Dalsgaard *et al.*, 1996). In the limit of large-scale source size ($d \ll L$), $f_{\mathrm{geom}} = 1/2$, and there is a wide range of nonthermal electron energy fluxes which yield pressure increases which could be comparable or even greater than the above Lorentz force example. On the other hand, for a small 1000 km flare kernel size, we find temperature enhancements that are comparable to the candidate Lorentz-force value only for the very largest nonthermal electron energy fluxes we have considered. Regarding the nonthermal electron energy flux levels, we must point out that recent RHESSI and *Hinode* observations (Krucker *et al.*, 2011) of the white-light flare of 6 December 2006 indicate a value of F_{nte} of $10^{12} - 10^{13}\ \mathrm{erg\,cm^{-2}\,s^{-1}}$, a value that greatly exceeds our assumed energy flux range in Figure 6, and which would result in a pressure increase that greatly exceeds the Lorentz-force estimate in that figure, even for a small value of L/d. However, we also point out an observed sunquake (15 February 2011) for which no white-light enhancement was observed at all (C.A. Lindsey and J.C. Martínez-Oliveras, 2011, private communication), indicating that in this case backwarming did not play a significant role.

We must caution that our perturbed-pressure estimates for backwarming are probably overestimates. Our treatment assumes that the temperature changes instantaneously (ignoring the time lag due to the finite heat capacity of the photospheric plasma), and it assumes that the photospheric plasma is frozen in place and does not respond dynamically to the increased heating (the plasma should expand in response to the enhanced heating on a sound-crossing time – for a 70 km thick photospheric layer, with $C_s \approx 8$ km s^{-1}, this is $\approx$ ten seconds). This treatment also ignores the possibility of multi-step radiative reprocessing, in which the backwarming radiation that reaches the photosphere comes not from the primary source in the flare chromosphere, but from secondary backheating sources, where the UV line emission is first converted via backwarming to other radiation mechanisms (*e.g.*, H bound–free continuum emission), before finally reaching the photosphere. Each step of a multi-step reprocessing will result in further dilution of the energy flux reaching the photosphere.

In summary, our simple estimate of backwarming-induced temperature and pressure increases shows a wide range of possible outcomes. To be competitive with the Lorentz-force surface density taken from Hudson, Fisher, and Welsch (2008) requires either backwarming sources that are much wider than their height above the photosphere, or for small flare kernel sizes, energy fluxes well in excess of 10^{11} erg cm^{-2} s^{-1}. Our pressure enhancement estimates are probably upper limits, since they ignore heat capacity effects, expansion of the heated photospheric plasma, and any secondary reprocessing. Nevertheless, these estimates provide useful guidelines for future, more detailed investigations of flare-driven backwarming.

4. Conclusions

We derive an expression for the vertical and horizontal components of the Lorentz-force change implied by observed magnetic-field changes occurring over the course of a solar flare. The Lorentz-force change acting on the outer solar atmosphere (Equations (9) – (10)) is balanced exactly by a corresponding Lorentz-force change acting on the photosphere and below: Equations (17) – (18). The Lorentz-force change, integrated over the time period over which the change occurs, defines an impulse. The impulse defines a momentum increase, given in Equations (14) and (15). The radial component of the impulse, acting on the outer solar atmosphere, is then used to derive an upper limit to the mass of CME ejecta that escape from the Sun: Equation (16).

We show that our expression for the vertical Lorentz-force change acting on the solar interior (Equation (17)) generalizes our earlier result in Hudson, Fisher, and Welsch (2008), in that it includes horizontal as well as vertical (radial) forces. It is also more accurate, in that it does not assume a first-order expansion of changes in the magnetic field.

The balance between the Lorentz forces acting on the solar atmosphere and the solar interior leads us to suggest a possible connection between the upward momentum in flare ejecta and the downward momentum in the solar interior, and leads to the possibility of using helioseismic measurements of "sunquakes" to study the properties of eruptive flares and CMEs.

To further elucidate the physical origins of "sunquake" acoustic emission, we also estimate force perturbations in the photosphere due to changes in gas pressure driven by chromospheric evaporation and from radiative backwarming of the photosphere during solar flares. We find, for chromospheric evaporation in flares, an upper limit of ≈ 2000 dyne cm^{-2} for a pressure increase in the upper chromosphere. We show that this pressure increase will lead to the downward propagation of a chromospheric condensation (a dense region behind a downward-moving shock-like disturbance), but the chromospheric condensation can propagate to column depths of at most a few percent of the photospheric depth; the subsequent propagation to the photosphere occurs by means of an acoustic disturbance. Whether this acoustic disturbance is significant at photospheric depths, when compared to the Lorentz force per unit area, is not yet clear, and will require a more detailed analysis of acoustic-wave propagation and dissipation effects in the solar chromosphere. We compare the Lorentz force to gas-pressure changes driven by radiative backwarming, and find the latter mechanism could be comparable to, or greater than, the Lorentz force if the region being energized by flare nonthermal electron heating has a horizontal extent much greater than its height above the photosphere, or for smaller heated regions, if the nonthermal electron energy flux greatly exceeds 10^{11} erg cm^{-2} s^{-1}. We caution that our estimates of pressure changes due to backheating are probably overestimates.

To summarize, the primary source for energy release in eruptive solar flares is most likely the solar magnetic field in strong-field, low-β active regions. It then makes sense that changes in the magnetic field itself will have a more direct and larger impact on the atmosphere than changes that are due to secondary flare processes, such as the production of energetic particles, gasdynamic motions, and enhanced radiative output, all of which are assumed to be driven ultimately by the release of magnetic energy.

Acknowledgements   This work was supported by the NASA Heliophysics Theory Program (grants NASA-NNX08AI56G and NASA-NNX11AJ65G), by the NASA LWS TR&T program (grants NNX08AQ30G and NNX11AQ56G), by the NSF SHINE program (grants ATM0551084 and ATM-0752597), by the NSF core AGS program (grant ATM0641303) funding our efforts in support of the University of Michigan's CCHM project, NSF NSWP grant AGS-1024682, and NSF core grant AGS-1048318. HSH acknowledges support from NASA Contract NAS5-98033. We thank the US taxpayers for making this research possible. We wish to acknowledge the role that Dick Canfield has played in realizing the importance of global momentum balance in the dynamics of solar-flare plasmas, through several seminal papers on this topic in the 1980s and 1990s. We thank the referee, Charlie Lindsey, for noting an error in our initial discussion of the depth dependence of the gas-pressure perturbation for downward-propagating acoustic waves. We also thank him for providing an exceptionally thoughtful, thorough, and detailed critique of this article, which improved it greatly.

References

Abbett, W.P., Hawley, S.L.: 1999, Dynamic models of optical emission in impulsive solar flares. *Astrophys. J.* **521**, 906–919. doi:10.1086/307576.

Allred, J.C., Hawley, S.L., Abbett, W.P., Carlsson, M.: 2005, Radiative hydrodynamic models of the optical and ultraviolet emission from solar flares. *Astrophys. J.* **630**, 573–586. doi:10.1086/431751.

Brown, J.C.: 1971, The deduction of energy spectra of non-thermal electrons in flares from the observed dynamic spectra of hard X-ray bursts. *Solar Phys.* **18**, 489–502. doi:10.1007/BF00149070.

Chen, Q.R., Ding, M.D.: 2005, On the relationship between the continuum enhancement and hard X-ray emission in a white-light flare. *Astrophys. J.* **618**, 537–542. doi:10.1086/425856.

Christensen-Dalsgaard, J., Dappen, W., Ajukov, S.V., Anderson, E.R., Antia, H.M., Basu, S., Baturin, V.A., Berthomieu, G., Chaboyer, B., Chitre, S.M., Cox, A.N., Demarque, P., Donatowicz, J., Dziembowski, W.A., Gabriel, M., Gough, D.O., Guenther, D.B., Guzik, J.A., Harvey, J.W., Hill, F., Houdek, G., Iglesias, C.A., Kosovichev, A.G., Leibacher, J.W., Morel, P., Proffitt, C.R., Provost, J., Reiter, J., Rhodes, E.J. Jr., Rogers, F.J., Roxburgh, I.W., Thompson, M.J., Ulrich, R.K.: 1996, The current state of solar modeling. *Science* **272**, 1286–1292. doi:10.1126/science.272.5266.1286.

Donea, A., Besliu-Ionescu, D., Cally, P.S., Lindsey, C., Zharkova, V.V.: 2006, Seismic emission from A M9.5-class solar flare. *Solar Phys.* **239**, 113 – 135. doi:10.1007/s11207-006-0108-3.

Fisher, G.H.: 1987, Explosive evaporation in solar flares. *Astrophys. J.* **317**, 502 – 513. doi:10.1086/165294.

Fisher, G.H.: 1989, Dynamics of flare-driven chromospheric condensations. *Astrophys. J.* **346**, 1019 – 1029. doi:10.1086/168084.

Fisher, G.H., Canfield, R.C., McClymont, A.N.: 1985a, Flare loop radiative hydrodynamics – part seven – dynamics of the thick target heated chromosphere. *Astrophys. J.* **289**, 434 – 441. doi:10.1086/162903.

Fisher, G.H., Canfield, R.C., McClymont, A.N.: 1985b, Flare loop radiative hydrodynamics – part six – chromospheric evaporation due to heating by nonthermal electrons. *Astrophys. J.* **289**, 425 – 433. doi:10.1086/162902.

Fisher, G.H., Canfield, R.C., McClymont, A.N.: 1985c, Flare loop radiative hydrodynamics. V – response to thick-target heating. *Astrophys. J.* **289**, 414 – 424. doi:10.1086/162901.

Fletcher, L., Hannah, I.G., Hudson, H.S., Metcalf, T.R.: 2007, A TRACE white light and RHESSI hard X-ray study of flare energetics. *Astrophys. J.* **656**, 1187 – 1196. doi:10.1086/510446.

Forbes, T.G.: 2000, A review on the genesis of coronal mass ejections. *J. Geophys. Res.* **105**, 23153 – 23166.

Hudson, H.S.: 2007, Chromospheric flares. In: Heinzel, P., Dorotovič, I., Rutten, R.J. (eds.) *The Physics of Chromospheric Plasmas* **CS-368**, Astron. Soc. Pacific, San Francisco, 365.

Hudson, H.S., Fisher, G.H., Welsch, B.T.: 2008, Flare energy and magnetic field variations. In: Howe, R., Komm, R.W., Balasubramaniam, K.S., Petrie, G.J.D. (eds.) *Subsurface and Atmospheric Influences on Solar Activity* **CS-383**, Astron. Soc. Pacific, San Francisco, 221 – 226.

Hudson, H.S., Acton, L.W., Hirayama, T., Uchida, Y.: 1992, White-light flares observed by YOHKOH. *Publ. Astron. Soc. Japan* **44**, L77 – L81.

Hudson, H.S., Fletcher, L., Fisher, G.H., Abbett, W.P., Russell, A.: 2011, Momentum distribution in solar flare processes. *Solar Phys.* **340**. doi:10.1007/s11207-011-9836-0.

Kosovichev, A.G.: 2011, Helioseismic response to the X2.2 solar flare of 2011 February 15. *Astrophys. J. Lett.* **734**, L15 – L20. doi:10.1088/2041-8205/734/1/L15.

Kosovichev, A.G., Zharkova, V.V.: 1995, Seismic response to solar flares: theoretical predictions. In: Hoeksema, J.T., Domingo, V., Fleck, B., Battrick, B. (eds.) *Helioseismology* **SP-376**, ESA, Noordwijk, 341 – 344.

Kosovichev, A.G., Zharkova, V.V.: 1998, X-ray flare sparks quake inside Sun. *Nature* **393**, 317 – 318. doi:10.1038/30629.

Krucker, S., Hudson, H.S., Jeffrey, N.L.S., Battaglia, M., Kontar, E.P., Benz, A.O., Csillaghy, A., Lin, R.P.: 2011, High-resolution imaging of solar flare ribbons and its implication on the thick-target beam model. *Astrophys. J.* **739**, 96. doi:10.1088/0004-637X/739/2/96.

Lin, R.P., Hudson, H.S.: 1976, Non-thermal processes in large solar flares. *Solar Phys.* **50**, 153 – 178. doi:10.1007/BF00206199.

Lindsey, C., Donea, A.: 2008, Mechanics of seismic emission from solar flares. *Solar Phys.* **251**, 627 – 639. doi:10.1007/s11207-008-9140-9.

Low, B.C.: 1985, Modeling solar magnetic structures. In: Hagyard, M.J. (ed.) *Measurements of Solar Vector Magnetic Fields*, NASA, Huntsville, 49 – 65.

Metcalf, T.R., Jiao, L., McClymont, A.N., Canfield, R.C., Uitenbroek, H.: 1995, Is the solar chromospheric magnetic field force-free? *Astrophys. J.* **439**, 474 – 481.

Metcalf, T.R., Alexander, D., Hudson, H.S., Longcope, D.W.: 2003, TRACE and Yohkoh observations of a white-light flare. *Astrophys. J.* **595**, 483 – 492. doi:10.1086/377217.

Moradi, H., Donea, A.C., Lindsey, C., Besliu-Ionescu, D., Cally, P.S.: 2007, Helioseismic analysis of the solar flare-induced sunquake of 2005 January 15. *Mon. Not. Roy. Astron. Soc.* **374**, 1155 – 1163. doi:10.1111/j.1365-2966.2006.11234.x.

Petrie, G.J.D., Sudol, J.J.: 2010, Abrupt longitudinal magnetic field changes in flaring active regions. *Astrophys. J.* **724**, 1218 – 1237. doi:10.1088/0004-637X/724/2/1218.

Sudol, J.J., Harvey, J.W.: 2005, Longitudinal magnetic field changes accompanying solar flares. *Astrophys. J.* **635**, 647 – 658. doi:10.1086/497361.

Tamres, D.H., Canfield, R.C., McClymont, A.N.: 1986, Beam-induced pressure gradients in the early phase of proton-heated solar flares. *Astrophys. J.* **309**, 409 – 420. doi:10.1086/164613.

Vernazza, J.E., Avrett, E.H., Loeser, R.: 1981, Structure of the solar chromosphere. III – Models of the EUV brightness components of the quiet-Sun. *Astrophys. J. Suppl.* **45**, 635 – 725. doi:10.1086/190731.

Wang, H., Liu, C.: 2010, Observational evidence of back reaction on the solar surface associated with coronal magnetic restructuring in solar eruptions. *Astrophys. J. Lett.* **716**, L195 – L199. doi:10.1088/2041-8205/716/2/L195.

Watanabe, K., Krucker, S., Hudson, H., Shimizu, T., Masuda, S., Ichimoto, K.: 2010, G-band and hard X-ray emissions of the 2006 December 14 flare observed by Hinode/SOT and Rhessi. *Astrophys. J.* **715**, 651 – 655. doi:10.1088/0004-637X/715/1/651.

Solar Phys (2012) 277:77–88
DOI 10.1007/s11207-011-9836-0

Momentum Distribution in Solar Flare Processes

**H.S. Hudson · L. Fletcher · G.H. Fisher · W.P. Abbett ·
A. Russell**

Received: 29 December 2010 / Accepted: 4 July 2011 / Published online: 4 October 2011
© Springer Science+Business Media B.V. 2011

Abstract We discuss the consequences of momentum conservation in processes related to
solar flares and coronal mass ejections (CMEs), in particular describing the relative importance of vertical impulses that could contribute to the excitation of seismic waves ("sunquakes"). The initial impulse associated with the primary flare energy transport in the impulsive phase contains sufficient momentum, as do the impulses associated with the acceleration of the evaporation flow (the chromospheric shock) or the CME itself. We note that the
deceleration of the evaporative flow, as coronal closed fields arrest it, will tend to produce
an opposite impulse, reducing the energy coupling into the interior. The actual mechanism
of the coupling remains unclear at present.

Keywords Solar flare

1. Introduction

The conservation of linear momentum has not often been considered in discussions of the
dynamics of solar flares and coronal mass ejections (CMEs). The exception to this is in the
evaporative flow, where several authors have described the theoretical (Brown and Craig,
1984; McClymont and Canfield, 1984) and observational (Zarro *et al.*, 1988; Canfield *et al.*,
1990) consequences: redshifts must occur to compensate for blueshifts as the chromosphere
expands. Indeed, recent spectroscopic observations have shown an interesting temperature
dependence of these red and blue shifts (*e.g.*, Milligan and Dennis, 2009), with a division at
about 2×10^6 K.

Solar Flare Magnetic Fields and Plasmas
Guest Editors: Y. Fan and G.H. Fisher

H.S. Hudson (✉) · G.H. Fisher · W.P. Abbett
SSL/University of California, Berkeley, CA, USA
e-mail: hhudson@ssl.berkeley.edu

H.S. Hudson · L. Fletcher · A. Russell
School of Physics and Astronomy, SUPA, University of Glasgow, Glasgow G12 8QQ, UK

In this article we qualitatively explore the consequences of momentum conservation in other aspects of solar flares. These include not only the momentum associated with the bodily transfer of mass, as with the evaporative flow and with CMEs, but also that represented by significant wave transport of energy (*e.g.*, Fletcher and Hudson, 2008; Haerendel, 2009). In fact, the low plasma β of the corona (*e.g.*, Gary, 2001) means that the momentum will reside mostly in the electromagnetic field, rather than in the matter. Energy transport *via* the Alfvénic Poynting flux (for a discussion in the context of magnetic-reconnection flare models see Birn *et al.*, 2009) must happen if a flare represents the release and redistribution of coronal energy storage, and its dissipation as chromospheric radiation.

Quantitative estimates of the impulse in the energy-release phase depend on our knowledge of the coronal magnetic field and the exact nature of its restructuring, and the skimpiness of this knowledge probably accounts for the lack of prior work on this subject. Solar flares occur in a complicated magnetized plasma environment often described in the approximation of ideal MHD. In principle, MHD simulations can explore the properties of momentum in flares and CMEs, but in practice this aspect of the physics is not emphasized. Simple arguments based on body forces acting on discrete objects (where does one push on a CME exactly?) generally are of less value than descriptions of the hydrodynamic aspects of the flows (see the description by Fisher *et al.*, 2011). Note that flare plasmas involve substantial particle acceleration that also must be included in momentum assessments (Brown and Craig, 1984; McClymont and Canfield, 1984). This aspect of the momentum balance would not be a part of any ideal MHD theory or simulation.

To a good approximation, a flare–CME occurs in a stationary solar atmosphere with zero net momentum. At the end of the process, if no CME has happened, another similar stationary state will result, although mass and energy will have been redistributed. If a CME does happen, mass and waves flow into the solar wind and are lost to the Sun forever, and this will also result in a displacement and a small change of the momentum of the body of the Sun. Here "small" can be put in the context that $\Delta v_\odot = m_{\mathrm{CME}}/M_\odot \times v_{\mathrm{CME}}$, of order 10^{-10} cm s^{-1}. This is doubtless entirely irrelevant for a solar-type star.

Flare seismic signatures in the solar interior ("sunquakes": Kosovichev and Zharkova, 1998) require momentum acquired from the coronal–chromospheric dynamics of a flare (Wolff, 1972). Zharkova and Zharkov (2007) and Zharkova (2008) discuss this problem in detail *via* analyses of the flares SOL1996-07-09T09:11 and SOL2003-10-28T11:10. Various mechanisms have been invoked to relate the seismic waves to the flare processes themselves, and the observations point to the flare footpoints during the impulsive phase (Kosovichev and Zharkova, 1998; Donea and Lindsey, 2005) as the seismic sources. Another characteristic of the impulsive phase is the evaporation flow that fills the coronal flare loops with hot plasma, creating the coronal X-ray sources. We point out (Section 2.2) that evaporation into closed fields implies a pair of impulses – a first impulse to accelerate the mass up into the corona, and a second and opposite one to arrest its motion there. We discuss the implications of this characteristic for seismic waves specifically in Section 3.

2. Application of Linear Momentum Conservation

2.1. Reference Flare Parameters

This article discusses momentum conservation in flares and CMEs. We only consider the vertical component of linear momentum. For a concrete context we consider a typical X1-class solar flare with a CME, and assume the parameters listed in Table 1. At this flare

Table 1 Representative parameters for an X-class flare with CME and sunquake.

Property	Value
Total energy of flare	10^{32} erg
Flare-loop height	1×10^9 cm
Coronal density (preflare)	1×10^9 cm^{-3}
Coronal field	1×10^3 G
Impulsive sub-burst duration	10 s
Impulsive-phase duration	100 s
Number of sub-bursts	10
Impulsive sub-burst footpoint area	3×10^{17} cm^2
Evaporation speed	5×10^7 cm s^{-1}
Evaporated mass	1×10^{14} g
Draining time	1000 s
CME mass	1×10^{15} g
CME speed	2×10^8 cm s^{-1}
Seismic-wave energy[a]	4×10^{27} erg

[a] Moradi *et al.* (2007).

magnitude, a CME is likely but sometimes does not happen; for less energetic flares, CME occurrence becomes less probable (Yashiro *et al.*, 2005; Wang and Zhang, 2007). Section 2.4 discusses the case of a flare with no CME.

As a guide to representative parameters of a flare–CME system, we assume that the flare impulsive phase consists of a series of ten independent impulsive sub-bursts as indicated in Table 1; this is just illustrative since a broad distribution of time scales for sub-bursts exists, ranging down to time scales below one second (Kiplinger *et al.*, 1984). The conceptual flare also involves a seismic wave (sunquake) containing 4×10^{27} erg, taken as 0.01% of the total flare energy (Moradi *et al.*, 2007). The information in Table 1 is meant to be representative and is certainly incomplete in the sense that it omits various features. We include the seismic wave because of its interesting diagnostic relationship to momentum conservation.

In the scheme considered (Figure 1), energy originates in the corona and flows into the flare footpoints either as in the standard thick-target model of an electron beam, or via Alfvénic Poynting flux (Fletcher and Hudson, 2008). The energy released in the footpoints drives the evaporative flow, which is arrested in an arcade of magnetic field and eventually drains back into the chromosphere. The complementary momentum for the evaporation flow appears in a downward wave structure in the deeper atmosphere (Kostiuk and Pikel'ner, 1975). From one equilibrium state to the next, this scheme involves four major impulse pairs with balanced vertical momentum components: the impulse associated with the primary energy release in the corona (a–a in Figure 1), that involved in the chromospheric heating and evaporation (b–b), that associated with the arrested evaporative flow (c–c), and (for completeness) that associated with the drained material (d–d) impacting the chromosphere (Hyder, 1967). The balancing impulses may be separated in time via transport of energy and momentum through the plasma by flows, waves, or particles. The flare results from that portion of the primary energy release carried into the lower solar atmosphere.

2.1.1. Beams

In the generally accepted picture, the energy of a flare comes from magnetic energy storage in the strong magnetic fields of an active region, on a characteristic scale of 10^9 cm (here

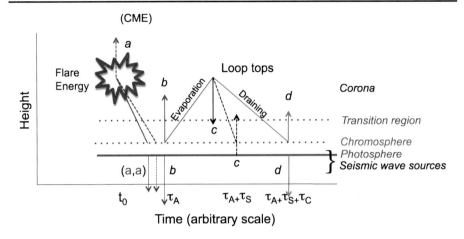

Figure 1 Timeline of the vertical impulses in an idealized solar flare with (optional) CME. The temporal axis is nonlinear and highly schematic in this representation. The initial primary energy release (the flare) communicates an impulse to the chromosphere on a time scale τ_a, reflecting either the beam propagation time (if particles convey the energy) or the Alfvén transport time (if waves). In either case this time scale is smaller than the evaporation time scale τ_s, which itself is shorter than the draining time scale τ_c. The impulsive-phase transport has two alternatives: the classical electron beam and the Poynting-flux alternative of Fletcher and Hudson (2008). The solid lines and arrows show energy transport and impulses in matter, and the dashed lines and arrows show the same in waves. The impulse pairs are labeled by letters, *i.e.* the pair *a–a* shows the pair related to primary energy release, *b–b* that associated with the flare heating itself, *etc.*

we restrict ourselves to events occurring in active regions). Timing evidence suggests that for CMEs associated with active-region flares, their energy too derives from a similar source (Dere *et al.*, 1997; Zarro *et al.*, 1999; Zhang *et al.*, 2004; Temmer *et al.*, 2008).

On the other hand, the radiated energy of a flare comes mainly from the chromosphere and photosphere (*e.g.*, Emslie *et al.*, 2005), and a substantial (if not dominant) part of this energy appears in the impulsive phase of the flare (here taken to be the rise phase of the GOES soft X-ray burst accompanying the flare). This means that the energy must propagate from its coronal storage region into the chromosphere on a relatively short time scale. The standard thick-target model (*e.g.*, Brown, 1971; Kane and Donnelly, 1971; Hudson, 1972) assigns this propagation to a beam of non-thermal electrons (see Section 2.1.2 for the Poynting-flux alternative). Variants of the thick-target model with protons (Najita and Orrall, 1970; Švestka, 1970) or neutral beams (Simnett and Haines, 1993) have also been proposed; these would contain larger momenta than the electron beams.

The vertical momentum transport by an electron beam in the thick-target model can be estimated from the observed hard X-ray flux (Brown and Craig, 1984; McClymont and Canfield, 1984). We can estimate the total momentum of the beam as $p = N m_e v_e$, where v_e is the mean vertical electron speed and N the total number of electrons. This omits several complicating factors, including the return current (Knight and Sturrock, 1977) required by the charge-neutrality condition, which could substantially reduce the momentum contained in the beam. Nevertheless if we generalize the model geometrically by allowing a curved flux tube, then the beam (and its anti-beam) will drive impulses (of the same sign) into the field (Section 2.1.2), in which case our simple estimate is of the right order of magnitude.

For a concrete example (one ten-second sub-burst) we take $E = 10^{31}$ erg and $v_e = 10^{10}$ cm s^{-1} for a momentum $p = 2 \times 10^{21}$ gm cm s^{-1}. The impulse imparted to the photosphere over an area A and time Δt corresponds to a beam pressure $P = p/A\Delta t =$

6×10^2 dyne cm^{-2} for beam area 3×10^{17} cm^2 and ten-second duration. Smaller times or areas would result in larger beam pressures. This illustrative number, based on TRACE and *Hinode* white-light flare observations (Hudson, Wolfson, and Metcalf, 2006; Fletcher *et al.*, 2007; Isobe *et al.*, 2007), makes an important point: the beam dynamic pressure exceeds the pressure of the ambient atmosphere, even in semi-empirical models of flare atmospheres such as the FLB model of Mauas, Machado, and Avrett (1990). In this flare model the pressure at $n = 10^{13}$ cm^{-3} is only about 20 dyne cm^{-2}. Brown and Craig (1984) also make this point about the radiative-transfer models, and McClymont and Canfield (1984) note that the hydrodynamic pressures due to heating and evaporation should be much greater in magnitude.

We conclude that the existing semi-empirical models of the lower atmosphere during a flare probably do not represent the impulsive phase well.

2.1.2. Waves

Energy transport via Alfvén waves at low plasma β implies a momentum flux S/v_A, where S is the (Alfvénic) Poynting flux, and v_A the Alfvén speed. This momentum flux, and the time scales of the reaction in the photosphere, are similar to those expected in the thick-target model. The high speeds of particles or Alfvén waves mean that only a small temporal interval separates the energy-release time in the low corona from the impulse applied to the Sun. This initial impulse begins to appear where the energy is absorbed; for the thick-target model this is normally calculated from the electron deflections by Coulomb scattering (Brown, 1971). In the case of wave transport, it depends upon the mechanism for wave damping, but this may ultimately be in the form of similar electron distributions (Fletcher and Hudson, 2008).

The Alfvén speed can be quite high in the core of an active region: 1000 G and $n_e = 10^9$ cm^{-3} corresponds to $v_A/c \approx 0.3$. Figure 2 (left) compares the ion sound speed with the Alfvén speed for Model 1006 (sunspot umbra) of Fontenla *et al.* (2009). Figure 2 (right) illustrates the slowing-down of an Alfvénic wave packet as it passes through the photosphere, using the same model. We have assumed a uniform magnetic field of 3000 G for this estimate, which leads to an elapsed time of 38 seconds between the top of the model and its base. At the base of this model atmosphere (164 km below $\tau_{5000} = 1$) the sound speed increases with depth, and energy deposited in this region can enter the interior and be trapped there as a sunquake. Fisher *et al.* (2011) discuss the theory of this coupling.

The case shown in Figure 2 is for a reasonable assumption about the magnetic field at the umbral photosphere. For the Poynting-flux model of the impulsive-phase energy transport (Fletcher and Hudson, 2008) we could interpret this roughly as the time delay between the hard X-ray burst and the injection time of acoustic energy into the interior. In principle, this delay would be different for particle or wave transport, and for direct photospheric heating via radiative backwarming (Machado, Emslie, and Avrett, 1989). Kosovichev (2007) discusses the complications resulting from multiple acoustic sources that may compete in a given flare event. In general, the timing delay seen in Figure 2 also suggests a filtering effect as regards sunquake amplitude, because it spreads the energy out temporally. For the case shown, the total delay $\tau \approx 38$ seconds would correspond to a wave frequency $f = (2\pi\tau)^{-1} \approx 5$ mHz, near the frequency band often used for sunquake studies, and the observations would not capture some fraction of the high-frequency power.

2.2. Evaporation

The evaporative flow presents a complicated set of momentum issues. The energy from the corona, by whatever means, heats the chromosphere impulsively and drives matter up

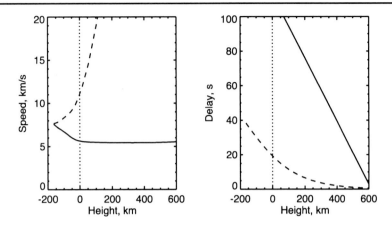

Figure 2 Left: sound- (solid) and Alfvén-speeds for the Fontenla *et al.* (2009) model atmosphere for a sunspot umbra (their Model 1006), assuming a magnetic field of 3000 G. Right: time elapsed for energy to arrive from the corona. The solid line shows the acoustic travel time, and the dashed line the Alfvénic.

and down. Momentum is conserved, and that of the upward mass flux (and waves) must be compensated by downward momentum. The upward flow is the chromospheric evaporation, and the downward flow converts into a shock wave that dissipates radiatively (Fisher *et al.*, 2011). This is the hydrodynamic response initially described by Kostiuk and Pikel'ner (1975). The momentum balance of the initial expansion can be observed via bisector analysis of chromospheric lines (Zarro *et al.*, 1988; Canfield *et al.*, 1990). The downward component eventually appears, as described, as a miniscule acceleration of the Sun, and some of its energy may excite sunquake acoustic waves (see Section 3) as envisioned by Kosovichev and Zharkova (1998).

The evaporated mass flows up into closed loops on relatively short time scales; at the loop top the vertical flow is arrested by the magnetic field, which provides the necessary impulse. This second (magnetic) impulse acts on the body of the Sun via wave coupling that we do not consider here; the general effect would be to launch a bipolar wave front (*i.e.*, two successive perturbations of opposite sign) into the solar interior. The second impulse would be spread out over a longer time scale owing to the dispersion of the evaporated mass as it flows up into the flux tube. The transfer of momentum into the photosphere will also be dispersed because of the wave propagation. Figure 1 sketches how the complete impulse might appear for a single ten-second pulse. This compensating impulse from the stoppage of the evaporation flow could presumably be exerted at the footpoints corresponding to the initial evaporation flow, or it could be spread more widely depending on the details of the wave transport of the momentum. The complicated properties of the momentum transfer associated with the presence of closed fields presumably will require analysis via numerical simulation, and we do not think that the standard 1D radiation hydrodynamics can capture the necessary physics since it omits the Lorentz force.

2.3. CMEs

Momentum balance in the CME ejection depends upon the time scale of the acceleration of the mass, plus unknown magnetic effects. The magnetic effects, as discussed above, must dominate initially if the energy source is in the low-β coronal field, but the momentum can appear ultimately in the mass flow swept up by the ejection; according to Table 1 the

Table 2 Vertical momentum components, representative X-class flare with CME.

Item Figure 1	Phenomenon	Mass g	v km s^{-1}	Δt s	Momentum g cm s^{-1}	Pressure dyne cm^{-2}
a	Primary (e$^-$)[a]	2×10^{11}	c/3	10	2×10^{21}	7×10^2
a	Primary (p$^+$ or H)[a]	1×10^{13}	2×10^8	10	1×10^{23}	3×10^5
a'	Primary (waves)	–	c/3	10	1×10^{21}	3×10^2
b	Evaporation flow	10^{14}	500	10	2×10^{22}	2×10^3
b'	Radiation[b]	–	c	10	1×10^{19}	3
c	CME	10^{15}	2000	100	2×10^{23}	?
d	Draining	10^{15}	10	$\approx 10^4$	2×10^{21}	0.07
	Seismic wave[c]				8×10^{21}	

[a] 20 keV if e$^-$, 20 MeV if p$^+$ or H.

[b] White-light flare.

[c] Kosovichev and Zharkova (1998), adjusted to X1.

observed CME momentum is consistent with this idea. The total magnitude of the mass component of CME momentum, for a major CME of 10^{15} g at a speed of 2×10^3 km s^{-1} is 2×10^{23} g cm s^{-1}. The source of the CME mass and its acceleration cannot be determined very completely from coronagraphic observations, owing to the presence of the occulting edge, and generally must be described in terms of non-coronagraphic observations in X-rays or at other wavelengths (*e.g.*, Hudson and Cliver, 2001). Observations of X-ray dimming (Hudson and Webb, 1997) and coronagraphic height-*vs.*-time plots (Zhang *et al.*, 2004; Temmer *et al.*, 2008) clearly point to the impulsive phase of the flare for flare-associated CMEs. However, the data are not good enough to determine the properties of the acceleration on the fine time scales of the impulsive-phase variations. The mass of a CME also includes (and may be dominated by) the mass swept up from the corona itself during the eruption, but on time scales much longer than those of the impulsive phase. Moreover the footprint in the photosphere of the magnetic structures involved in CME formation is not understood in detail, and so we generally only have upper limits on the spatial and temporal scales of the impulse imparted to the photosphere. Hence Table 2 has no entry for the pressure that the CME produces in the lower atmosphere.

2.4. No CMEs

In general a flare does not have an accompanying CME, even for some X-class events (Wang and Zhang, 2007). In such a case, how can momentum be conserved in the initial energy release? We can only speculate about this, since there do not appear to be any relevant observations. We suggest that the upward impulse, needed to balance the downward push on the photosphere, takes the form of waves radiated upwards and not damped in the mass flow of the CME. In this situation the plasma has no bulk flow and the waves propagate relatively freely through the nearly stationary medium, without rapid damping (Axford and McKenzie, 1992). As described by Belcher (1971), the wave energy can eventually exert substantial pressure on the solar wind as the waves damp. Because we rarely see type II radio bursts in the absence of CMEs (but see Klein, Trottet, and Klassen, 2010 for a good example of one), we suspect that the magnetic-pressure pulse may be generally more gradual than a gas-pressure pulse would be, because of weak damping. This would presumably soften the wave front and delay the "ignition" of the type II emission because the shock condition would not be met so readily (*e.g.*, Vršnak and Lulić, 2000).

2.5. Summary

Figure 1 schematically summarizes the momentum transport in a flare–CME. The impulse associated with flare energy release in the corona, and delivered downwards, must appear ultimately in the photosphere. The recoil takes the form of the sunquake waves. The upward impulse may escape into the solar wind if a CME occurs, but the theory remains to be worked out. The stepwise changes observed in the photospheric field presumably are involved in the momentum transfer, but this will not be known quantitatively until these changes can be observed in the vector field. The propagation of the force between the corona and the photosphere presumably involves Alfvénic wave packets (Song and Lysak, 1994) with general motions of the plasma, including both tension and pressure forces. The waves may include non-MHD properties as well. We summarize the estimated momentum values in Table 2, which uses the representative flare parameters given in Table 1.

3. Seismic Waves

Flare-related seismic waves were first observed by Kosovichev and Zharkova (1998) following the prediction by Wolff (1972). Wolff also described the momentum transfer that we discuss here, and also the excitation of the p-mode standing waves. Observations have shown that the "sunquake" seismic waves originate in the impulsive phase, specifically the locations of the white-light flare, the magnetic transients, and the hard X-ray footpoints (Kosovichev and Zharkova, 1998; Donea and Lindsey, 2005). A sharp blow to the photosphere should excite a broad spectrum of acoustic waves, and the best observations of sunquakes come from frequencies above the p-mode power peak and into the $\sim f^{-2}$ high-frequency variability spectrum of solar emission. Note that observations to date have been limited to image cadences of $\approx$ one minute, corresponding to $f \leq 8.33$ mHz.

The detailed physical mechanism of energy and momentum transfer from the corona into the solar interior remains ill-understood. Three basic mechanisms have been proposed: the essentially hydrodynamic shock-wave heating originating in the chromosphere (Kostiuk and Pikel'ner, 1975; Kosovichev and Zharkova, 1998), the $j \times B$ forces from the inevitable magnetic transient (Anwar *et al.*, 1993; Kosovichev and Zharkova, 2001; Sudol and Harvey, 2005; Hudson, Fisher, and Welsch, 2008), and photospheric backwarming (Machado, Emslie, and Avrett, 1989; Martínez-Oliveros, Moradi, and Donea, 2008). The requirement for momentum conservation can in principle help to distinguish among these plausible mechanisms.

The sketch in Figure 1 and the entries in Table 2 show which momentum components could couple well with the solar interior. In the table, the components a, a', b, and b' are all estimated for the sub-pulse quantities (b' is the reaction to the radiation pressure of the flare continuum emission, not shown in Figure 1). Entries for components c, d are for the entire flare–CME.

From the momentum point of view, within the accuracy of these estimates, the likeliest sources of the seismic wave would be components a or a' (beam or Poynting-flux transport), b (evaporation), or possibly c (the CME). Items a, a', and b' (radiation pressure) may be excludable because of relatively small momentum transport, but we need better data. Item *d* (the draining of the flare-loop system) would be on too long a time scale for the observed seismic waves, and this might be the case for the CME impulse as well. The entry in the table assumes that the entire mass of the CME is accelerated during the impulsive phase, which is certainly an overestimate. Much of the CME mass may come from higher altitudes (*e.g.*, Burkepile *et al.*, 2004) hence requiring long wave-propagation times to couple to the photosphere and a poorer match to the observed frequency range for the seismic waves. The overpressure created by these various impulses (the right-hand column of Table 2) again offers several possibilities, but this overpressure needs to be delivered at or below the photospheric level because of the temporal scales involved (see Figure 2). The ill-understood nature of the CME footprint in the lower solar atmosphere would also be a consideration; it seems likely that the impulse associated with the CME acceleration may press on an extended area of the photosphere, including regions with lower Alfvén speeds.

There is an important caveat regarding chromospheric heating (evaporation) as a source of momentum for seismic waves, as illustrated in Figure 1: the motion of the evaporated mass is arrested by closed fields, which indeed should be in the process of collapsing anyway (Švestka *et al.*, 1987; Hudson, 2000; Wang and Liu, 2010). This impulse tends to counteract the initial impulse of the explosion, producing a negative impulse as described above and shown in Figure 1. The separation between these impulses would be the time scale for the evaporative flow, which is limited by the ion sound speed. Observations at wave frequencies smaller than the inverse of this time scale would not detect so much seismic energy; currently a typical frequency range for seismic observations is $5-7$ mHz, which corresponds to a time scale of about 160 seconds. Furthermore, multiple elementary impulses (our example has ten seconds) would tend to overlap and confuse one another (Kosovichev, 2007). Thus we need to regard the momentum inferred from the flare explosion as an upper limit to what could be coupled into the solar interior.

4. Conclusions

The momentum available from flare dynamics appears to be adequate to couple energy into interior seismic waves; within likely uncertainties the momentum could be that associated with the primary energy release from the corona, the overpressure associated with the evaporative flow, or conceivably the CME acceleration. The white-light flare radiation itself contains insufficient momentum, although the backwarming it might induce could suffice in principle (*e.g.*, Moradi *et al.*, 2007). We have noted that the evaporative flow tends to be self-canceling since its motion is arrested, and this action produces an opposite impulse. This would tend to reduce the sunquake amplitudes at lower frequencies, but the time scales would depend on the detailed geometry of the flare. We suggest that an analysis of this impulse could be informative, in the sense that the seismic wave could provide quantitative information about the evaporation process. The time scales are indecisive at present because of the limitations, both inherent and practical, of the wave observations. The exact mechanisms involved with the linkage between solar exterior and solar interior remain unclear, although several plausible schemes have been proposed.

The seismic-wave impulses in principle test our knowledge of the solar interior at its interface with an active region. The coupling of energy between the exterior and the interior, on a specified time scale, depends sensitively on the structure of the atmosphere in the active region (see Figure 2). We may hope that comprehensive observations of the impulsive-phase signatures of solar flares, and their induced seismic waves, can help us to understand flare-induced perturbations of the solar atmosphere.

The requirement for momentum conservation at the point of initial energy release implies that comparable amounts of energy must be lost to sinks other than the chromospheric radiation or evaporation processes. If this momentum is not absorbed by the CME ejection, which could not be the case in a flare without a CME (*e.g.*, Klein, Trottet, and Klassen, 2010), it must be lost into the solar wind and could be detectable eventually by other means. Its presence requires an increase in the total energy of a flare over and above the amounts needed for the flare emission and the CME ejection, as considered in current estimates. It could appear in the lower atmosphere in a form too diffuse to have been detected yet, or more likely it could be hidden in the solar wind.

Acknowledgements   This work was supported by NASA under contract NAS 5-98033 for RHESSI. Support at Glasgow came from the EU's SOLAIRE Research and Training Network at the University of Glasgow (MTRN-CT-2006-035484), the EC-funded HESPE project (FP7-2010-SPACE-1-263086), Rolling Grant ST/F002637/1 from the UK's Science and Technology Facilities Council, Leverhulme grant F00-179A, and a Fellowship from the Royal Commission for the Exhibition of 1851 (AR).

References

Anwar, B., Acton, L.W., Hudson, H.S., Makita, M., McClymont, A.N., Tsuneta, S.: 1993, Rapid sunspot motion during a major solar flare. *Solar Phys.* **147**, 287 – 303. doi:10.1007/BF00690719.

Axford, W.I., McKenzie, J.F.: 1992, The origin of high speed solar wind streams. In: Marsch, E., Schwenn, R. (eds.) *Solar Wind Seven Colloquium*, Pergamon, Oxford, 1 – 5.

Belcher, J.W.: 1971, Alfvénic wave pressures and the solar wind. *Astrophys. J.* **168**, 509. doi:10.1086/151105.

Birn, J., Fletcher, L., Hesse, M., Neukirch, T.: 2009, Energy release and transfer in solar flares: Simulations of three-dimensional reconnection. *Astrophys. J.* **695**, 1151 – 1162. doi:10.1088/0004-637X/695/2/1151.

Brown, J.C.: 1971, The deduction of energy spectra of non-thermal electrons in flares from the observed dynamic spectra of hard X-ray bursts. *Solar Phys.* **18**, 489 – 502. doi:10.1007/BF00149070.

Brown, J.C., Craig, I.J.D.: 1984, The importance of particle beam momentum in beam-heated models of solar flares. *Astron. Astrophys.* **130**, L5 – L7.

Burkepile, J.T., Hundhausen, A.J., Stanger, A.L., St. Cyr, O.C., Seiden, J.A.: 2004, Role of projection effects on solar coronal mass ejection properties: 1. A study of CMEs associated with limb activity. *J. Geophys. Res.* **109**, A03103. doi:10.1029/2003JA010149.

Canfield, R.C., Metcalf, T.R., Zarro, D.M., Lemen, J.R.: 1990, Momentum balance in four solar flares. *Astrophys. J.* **348**, 333 – 340. doi:10.1086/168240.

Dere, K.P., Brueckner, G.E., Howard, R.A., Koomen, M.J., Korendyke, C.M., Kreplin, R.W., Michels, D.J., Moses, J.D., Moulton, N.E., Socker, D.G., St. Cyr, O.C., Delaboudinière, J.P., Artzner, G.E., Brunaud, J., Gabriel, A.H., Hochedez, J.F., Millier, F., Song, X.Y., Chauvineau, J.P., Marioge, J.P., Defise, J.M., Jamar, C., Rochus, P., Catura, R.C., Lemen, J.R., Gurman, J.B., Neupert, W., Clette, F., Cugnon, P., van Dessel, E.L., Lamy, P.L., Llebaria, A., Schwenn, R., Simnett, G.M.: 1997, EIT and LASCO observations of the initiation of a coronal mass ejection. *Solar Phys.* **175**, 601 – 612. doi:10.1023/A:1004907307376.

Donea, A.C., Lindsey, C.: 2005, Seismic emission from the solar flares of 2003 October 28 and 29. *Astrophys. J.* **630**, 1168 – 1183. doi:10.1086/432155.

Emslie, A.G., Dennis, B.R., Holman, G.D., Hudson, H.S.: 2005, Refinements to flare energy estimates: A followup to "Energy partition in two solar flare/CME events" by A.G. Emslie *et al. J. Geophys. Res.* **110**, 11103. doi:10.1029/2005JA011305.

Fisher, G.A., Bercik, D.J., Welsch, B.T., Hudson, H.S.: 2011, Momentum balance in eruptive solar flares: The vertical Lorentz force acting on the solar atmosphere and the solar interior. *Solar Phys.*, submitted.

Fletcher, L., Hudson, H.S.: 2008, Impulsive phase flare energy transport by large-scale Alfvén waves and the electron acceleration problem. *Astrophys. J.* **675**, 1645 – 1655. doi:10.1086/527044.

Fletcher, L., Hannah, I.G., Hudson, H.S., Metcalf, T.R.: 2007, A TRACE white light and RHESSI Hard X-ray study of flare energetics. *Astrophys. J.* **656**, 1187 – 1196. doi:10.1086/510446.

Fontenla, J.M., Curdt, W., Haberreiter, M., Harder, J., Tian, H.: 2009, Semiempirical models of the solar atmosphere. III. Set of non-LTE models for far-ultraviolet/extreme-ultraviolet irradiance computation. *Astrophys. J.* **707**, 482 – 502. doi:10.1088/0004-637X/707/1/482.

Gary, G.A.: 2001, Plasma beta above a solar active region: Rethinking the paradigm. *Solar Phys.* **203**, 71 – 86.

Haerendel, G.: 2009, Chromospheric evaporation via Alfvén waves. *Astrophys. J.* **707**, 903 – 915. doi:10.1088/0004-637X/707/2/903.

Hudson, H.S.: 1972, Thick-target processes and white-light flares. *Solar Phys.* **24**, 414 – 428.

Hudson, H.S.: 2000, Implosions in coronal transients. *Astrophys. J. Lett.* **531**, L75 – L77. doi:10.1086/312516.

Hudson, H.S., Cliver, E.W.: 2001, Observing coronal mass ejections without coronagraphs. *J. Geophys. Res.* **106**, 25199 – 25214. doi:10.1029/2000JA004026.

Hudson, H.S., Webb, D.F.: 1997, Soft X-Ray Signatures of Coronal Ejections. In: Crooker, N., Joselyn, J.A., Feynman, J. (eds.) *Coronal Mass Ejections, Geophys. Monogr.* **99**, 27 – 38.

Hudson, H.S., Fisher, G.H., Welsch, B.T.: 2008, Flare energy and magnetic field variations. In: Howe, R., Komm, R.W., Balasubramaniam, K.S., Petrie, G.J.D. (eds.) *Subsurface and Atmospheric Influences on Solar Activity* **CS-383**, Astron. Soc. Pacific, San Francisco, 221 – 226.

Hudson, H.S., Wolfson, C.J., Metcalf, T.R.: 2006, White-light flares: A TRACE/RHESSI overview. *Solar Phys.* **234**, 79 – 93. doi:10.1007/s11207-006-0056-y.

Hyder, C.L.: 1967, A phenomenological model for disparitions brusques followed by flarelike chromospheric brightenings, II: Observations in active regions. *Solar Phys.* **2**, 267 – 284. doi:10.1007/BF00147842.

Isobe, H., Kubo, M., Minoshima, T., Ichimoto, K., Katsukawa, Y., Tarbell, T.D., Tsuneta, S., Berger, T.E., Lites, B., Nagata, S., Shimizu, T., Shine, R.A., Suematsu, Y., Title, A.M.: 2007, Flare ribbons observed with G-band and Fe I 6302 Å, filters of the solar optical telescope on board Hinode. *Publ. Astron. Soc. Japan* **59**, S807 – S813.

Kane, S.R., Donnelly, R.F.: 1971, Impulsive hard X-ray and ultraviolet emission during solar flares. *Astrophys. J.* **164**, 151 – 163. doi:10.1086/150826.

Kiplinger, A.L., Dennis, B.R., Frost, K.J., Orwig, L.E.: 1984, Fast variations in high-energy X-rays from solar flares and their constraints on nonthermal models. *Astrophys. J. Lett.* **287**, L105 – L108. doi:10.1086/184408.

Klein, K., Trottet, G., Klassen, A.: 2010, Energetic particle acceleration and propagation in strong CME-less flares. *Solar Phys.* **263**, 185 – 208. doi:10.1007/s11207-010-9540-5.

Knight, J.W., Sturrock, P.A.: 1977, Reverse current in solar flares. *Astrophys. J.* **218**, 306 – 310. doi:10.1086/155683.

Kosovichev, A.G.: 2007, The cause of photospheric and helioseismic responses to solar flares: High-energy electrons or protons? *Astrophys. J. Lett.* **670**, L65 – L68. doi:10.1086/524036.

Kosovichev, A.G., Zharkova, V.V.: 1998, X-ray flare sparks quake inside Sun. *Nature* **393**, 317 – 318. doi:10.1038/30629.

Kosovichev, A.G., Zharkova, V.V.: 2001, Magnetic energy release and transients in the solar flare of 2000 July 14. *Astrophys. J. Lett.* **550**, L105 – L108. doi:10.1086/319484.

Kostiuk, N.D., Pikel'ner, S.B.: 1975, Gasdynamics of a flare region heated by a stream of high-velocity electrons. *Sov. Astron.* **18**, 590 – 599.

Machado, M.E., Emslie, A.G., Avrett, E.H.: 1989, Radiative backwarming in white-light flares. *Solar Phys.* **124**, 303 – 317. doi:10.1007/BF00156272.

Martínez-Oliveros, J.C., Moradi, H., Donea, A.: 2008, Seismic emissions from a highly impulsive M6.7 solar flare. *Solar Phys.* **251**, 613 – 626. doi:10.1007/s11207-008-9122-y.

Mauas, P.J.D., Machado, M.E., Avrett, E.H.: 1990, The white-light flare of 1982 June 15–models. *Astrophys. J.* **360**, 715 – 726. doi:10.1086/169157.

McClymont, A.N., Canfield, R.C.: 1984, The unimportance of beam momentum in electron-heated models of solar flares. *Astron. Astrophys.* **136**, L1 – L4.

Milligan, R.O., Dennis, B.R.: 2009, Velocity characteristics of evaporated plasma using Hinode/EUV imaging spectrometer. *Astrophys. J.* **699**, 968 – 975. doi:10.1088/0004-637X/699/2/968.

Moradi, H., Donea, A., Lindsey, C., Besliu-Ionescu, D., Cally, P.S.: 2007, Helioseismic analysis of the solar flare-induced sunquake of 2005 January 15. *Mon. Not. Roy. Astron. Soc.* **374**, 1155 – 1163. doi:10.1111/j.1365-2966.2006.11234.x.

Najita, K., Orrall, F.Q.: 1970, White light events as photospheric flares. *Solar Phys.* **15**, 176 – 194.

Simnett, G.M., Haines, M.G.: 1993, On the production of hard X-rays in solar flares. *Solar Phys.* **130**, 253 – 263.

Song, Y., Lysak, R.L.: 1994, Alfvénon, driven reconnection and the direct generation of the field-aligned current. *Geophys. Res. Lett.* **21**, 1755 – 1758. doi:10.1029/94GL01327.

Sudol, J.J., Harvey, J.W.: 2005, Longitudinal magnetic field changes accompanying solar flares. *Astrophys. J.* **635**, 647 – 658. doi:10.1086/497361.

Temmer, M., Veronig, A.M., Vršnak, B., Rybák, J., Gömöry, P., Stoiser, S., Maričić, D.: 2008, Acceleration in fast halo CMEs and synchronized flare HXR bursts. *Astrophys. J. Lett.* **673**, L95 – L98. doi:10.1086/527414.

Švestka, Z.F., Fontenla, J.M., Machado, M.E., Martin, S.F., Neidig, D.F.: 1987, Multi-thermal observations of newly formed loops in a dynamic flare. *Solar Phys.* **108**, 237–250. doi:10.1007/BF00214164.

Švestka, Z.: 1970, The phase of particle acceleration in the flare development. *Solar Phys.* **13**, 471–489.

Vršnak, B., Lulić, S.: 2000, Formation of coronal Mhd shock waves – I. The basic mechanism. *Solar Phys.* **196**, 157–180.

Wang, H., Liu, C.: 2010, Observational evidence of back reaction on the solar surface associated with coronal magnetic restructuring in solar eruptions. *Astrophys. J. Lett.* **716**, L195–L199. doi:10.1088/2041-8205/716/2/L195.

Wang, Y., Zhang, J.: 2007, A comparative study between eruptive X-class flares associated with coronal mass ejections and confined X-class flares. *Astrophys. J.* **665**, 1428–1438. doi:10.1086/519765.

Wolff, C.L.: 1972, Free oscillations of the Sun and their possible stimulation by solar flares. *Astrophys. J.* **176**, 833–842. doi:10.1086/151680.

Yashiro, S., Gopalswamy, N., Akiyama, S., Michalek, G., Howard, R.A.: 2005, Visibility of coronal mass ejections as a function of flare location and intensity. *J. Geophys. Res.* **110**, A12S051–A12S0511. doi:10.1029/2005JA011151.

Zarro, D.M., Canfield, R.C., Metcalf, T.R., Strong, K.T.: 1988, Explosive plasma flows in a solar flare. *Astrophys. J.* **324**, 582–589. doi:10.1086/165919.

Zarro, D.M., Sterling, A.C., Thompson, B.J., Hudson, H.S., Nitta, N.: 1999, SOHO EIT observations of extreme-ultraviolet "Dimming" associated with a halo coronal mass ejection. *Astrophys. J. Lett.* **520**, L139–L142. doi:10.1086/312150.

Zhang, J., Dere, K.P., Howard, R.A., Vourlidas, A.: 2004, A study of the kinematic evolution of coronal mass ejections. *Astrophys. J.* **604**, 420–432. doi:10.1086/381725.

Zharkova, V.V.: 2008, The mechanisms of particle kinetics and dynamics leading to seismic emission and sunquakes. *Solar Phys.* **251**, 665–666. doi:10.1007/s11207-008-9266-9.

Zharkova, V.V., Zharkov, S.I.: 2007, On the origin of three seismic sources in the proton-rich flare of 2003 October 28. *Astrophys. J.* **664**, 573–585. doi:10.1086/518731.

Solar Phys (2012) 277:89–118
DOI 10.1007/s11207-011-9821-7

Modeling and Interpreting the Effects of Spatial Resolution on Solar Magnetic Field Maps

K.D. Leka · G. Barnes

Received: 4 April 2011 / Accepted: 20 June 2011 / Published online: 27 July 2011
© Springer Science+Business Media B.V. 2011

Abstract Different methods for simulating the effects of spatial resolution on magnetic field maps are compared, including those commonly used for inter-instrument comparisons. The investigation first uses synthetic data, and the results are confirmed with *Hinode*/SpectroPolarimeter data. Four methods are examined, one which manipulates the Stokes spectra to simulate spatial-resolution degradation, and three "post-facto" methods where the magnetic field maps are manipulated directly. Throughout, statistical comparisons of the degraded maps with the originals serve to quantify the outcomes. Overall, we find that areas with inferred magnetic fill fractions close to unity may be insensitive to optical spatial resolution; areas of sub-unity fill fractions are very sensitive. Trends with worsening spatial resolution can include increased average field strength, lower total flux, and a field vector oriented closer to the line of sight. Further-derived quantities such as vertical current density show variations even in areas of high average magnetic fill fraction. In short, unresolved maps fail to represent the distribution of the underlying unresolved fields, and the "post-facto" methods generally do not reproduce the effects of a smaller telescope aperture. It is argued that selecting a method in order to reconcile disparate spatial resolution effects should depend on the goal, as one method may better preserve the field distribution, while another can reproduce spatial resolution degradation. The results presented should help direct future inter-instrument comparisons.

Keywords Active regions, magnetic fields · Active regions, models · Instrumental effects · Magnetic fields, photosphere · Polarization, optical

Solar Flare Magnetic Fields and Plasmas
Guest Editors: Y. Fan and G.H. Fisher

K.D. Leka (✉) · G. Barnes
CoRA Division, NorthWest Research Associates, 3380 Mitchell Ln., Boulder, CO 80301, USA
e-mail: leka@cora.nwra.com

G. Barnes
e-mail: graham@cora.nwra.com

1. Introduction

Understanding the limits of the data used to analyze and interpret the state of a system is a necessary part of remote-sensing science. For more than a century, the Zeeman effect in magnetically sensitive spectral lines has been used to detect and interpret the presence and character of solar magnetic fields. Much of solar physics research relies on interpreting magnetic field "maps" to investigate the physical state and dynamical evolution of the solar plasma. Quantities such as the magnetic field strength and direction, its variation (gradient) with space and time, the current density (or magnetic twist, current helicity, or shear angles, as preferred), plasma velocity vector inferred in part from the Doppler signal of the polarization spectra, and a variety of magnetic-related forces and torques are all of interest. They form the basis for our understanding of active region structure, large-scale field structure – even the dynamo(s), corona, and solar wind production. And they are all available from these measurements of the solar magnetic field, or are they?

With advancing capability of detector technology, modulator design and larger photon-gathering capabilities, it has become a challenge to reconcile the differing results from different instruments that engage different observing schemes, using different optical layouts and telescope sizes.

Comparison efforts between instruments and their resulting magnetic field maps are not new. Considerable effort has gone into comparisons between observing programs which produce the line-of-sight component over the whole solar disk (*e.g.*, Tran *et al.*, 2005; Demidov *et al.*, 2008; Demidov and Balthasar, 2009), as these data products provide input to heliospheric models which are the center of both ongoing research and real-time space-weather applications. Line selection and spectral sampling are crucial to consider for comparisons when the instruments and final data products may appear quite similar (Ulrich *et al.*, 2002, 2009). A challenging task is to compare instruments whose observing approaches are very different, as in the comparisons between the scanning-slit Advanced Stokes Polarimeter (ASP) and the filter-based SOUP instrument (Berger and Lites, 2002), the ASP and MDI (Berger and Lites, 2003), *Hinode*/SP and MDI (Moon *et al.*, 2007), and the ASP and the Imaging Vector Magnetograph (Labonte, Mickey, and Leka, 1999). The latter comparison attempted to evaluate the performance of two vector-field data sources, which means including the additional complications of the linear polarization and its data products (the component of the field perpendicular, or transverse to, the line of sight, and its azimuthal angle) in addition to the circular polarization and line-of-sight magnetic field component. Such an effort is not new (Wang *et al.*, 1992; Varsik, 1995; Bao *et al.*, 2000; Zhang *et al.*, 2003), and the effort required has not become simpler with time.

The spatial-resolution issue is the focus here. It has come to our attention, primarily through renewed efforts to inter-compare the performance of different facilities (the "Vector Magnetic Field Comparison Group", an ad hoc group of which the authors are members, that the manner in which different instrumental resolutions are incorporated into these comparisons can lead to erroneous results, in the direction of false confidence – implying that there is little or no impact to the resulting data due to spatial resolution, when we argue here that this is not the case.

Below we describe a way to model the gross effects from instrumental spatial resolution for spectro-polarimetric data, and demonstrate how this is required in order to avoid misleading results from *post facto* re-binning ("post-facto" here meaning "applied after the inversion from spectra to field", such that it is the magnetogram itself which is "rebinned"). We demonstrate, using both synthetic and real data, that spatial resolution differences do in fact lead to different results. On a positive note, in some cases the effects of varying spatial resolution behave in a predictable and systematic manner that depends on the structure of the observed solar feature, a result which can guide the interpretation of data obtained at any given spatial resolution.

2. Demonstration: Real Data

We begin with an example of the issue: we want to use data from two instruments interchangeably, so how do they compare? As an example, we take NOAA Active Region 10953 observed on 30 April 2007. For this date, there exist co-temporal data from both the Michelson Doppler Interferometer ("MDI") aboard the *Solar and Heliospheric Observatory* ("SoHO", Scherrer *et al.*, 1995), and from the Solar Optical Telescope/SpectroPolarimeter aboard the *Hinode* mission (Kosugi *et al.*, 2007; Tsuneta *et al.*, 2008); these exact data were used in De Rosa *et al.* (2009) as a boundary condition for nonlinear force-free extrapolations. The level-1.8.1 MDI "Full-Disk Magnetogram" from 22:24 UT 30 April 2007 samples with 1.98″ at SoHO's L-1 location, which matches the optical spatial resolution of the telescope. The *Hinode*/SP scan which began at 22:30 UT 30 April 2007[1] is a "fast scan" which performs on-chip summation for the sub-critically sampled data, providing a final 0.3″-sampled map that effectively matches the telescope resolution. The MDI and *Hinode*/SP maps are shown in Figure 1, where B_{los}, the line-of-sight component of the "pixel-area averaged" field is used for the *Hinode*/SP vector magnetogram to ensure a consistent comparison with the MDI map, where the fill fraction is assumed unity throughout. (For reference, a brief table of terminology used herein is included with Table 1.)

A sub-region of the MDI data is selected to match the *Hinode*/SP field of view, to within a fraction of an MDI pixel. The total of the unsigned data is computed (Table 2) at the original spatial sampling. We then "sampled" the *Hinode*/SP B_{los} map using the IDL "congrid" routine and recompute the total of the unsigned result. No further checks are made on the inter-instrument calibration. We explicitly do not quote uncertainties at this point: the uncertainties for the sums are significantly smaller than the differences between the compared data sets, and even the effect of a bias due to different photon noise levels is not significant in this case.

Why is there a difference between results from *Hinode*/SP and MDI? With studies showing that MDI generally underestimates the line-of-sight signal (Berger and Lites, 2003; Tran *et al.*, 2005; Ulrich *et al.*, 2009), it seems contradictory that the MDI result is the larger (see Appendix B). Some difference can be attributed to the different lines used and the different heights thus sampled (*e.g.*, Ulrich *et al.*, 2009), and the different inversion methods employed. Naively (or rhetorically) assuming that these differences are accounted for, the obvious remaining factor is the spatial resolution between the two datasets. Worse spatial resolution is expected to dilute a polarization signal (Leka, 1999; Orozco Suárez *et al.*, 2007); if this is the case, why is there only a tiny difference between the two "resolutions" of the *Hinode* data when rebinned in this manner?

3. Demonstration: Synthetic Data

Light entering a polarimeter is partially polarized, with the fraction and direction of polarization a function of many things including the strength and direction of the magnetic field along the photon ray-path above the photospheric $\tau = 1$ layer. Light entering a telescope

---

[1] Inversion from level-1D spectra to a magnetic map courtesy Dr. B.W. Lites, using the HAO Milne-Eddington inversion code (Skumanich and Lites, 1987) modified for *Hinode*/SP data, and presented to the authors for use in De Rosa *et al.* (2009).

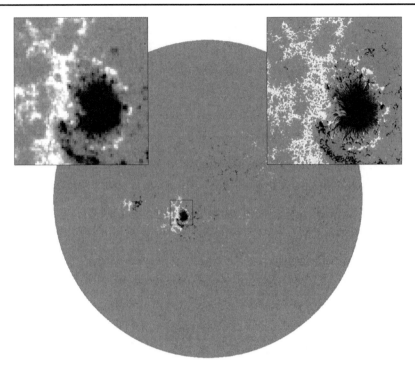

Figure 1 Data from the Michelson Doppler Imager (full-disk) line-of-sight component of the "pixel-area averaged" field B_{los} at 22:24 UT, 30 April 2007 includes NOAA Active Region 10953, delineated by a box. This area is also shown magnified (left inset). The same quantity for the same area on the Sun, derived from a Milne–Eddington inversion of *Hinode*/SpectroPolarimeter data obtained during 22:30 – 23:15 UT, 30 April 2007 is shown (right inset); all images are saturated at ± 500 Mx cm^{-2} (Mx: maxwell).

Table 1 Table of Magnetic Field Terminology.

Term	Symbol (if appropriate)	Meaning
Field strength	B	Magnitude of the field (given in G (gauss))
Fill fraction	f	Fraction of a pixel filled with field
Inclination angle	γ	Inclination to the line of sight 0°, 180° along the line of sight, 90° in the plane of the sky
Azimuthal angle	ϕ	Azimuthal angle
"Pixel-area averaged"		Either $f = 1.0$ is assumed, or the inferred fill fraction has been multiplied through (given in Mx cm^{-2}).
Line-of-sight component	B_{los}	$f B \times \cos(\gamma)$
Transverse component	B_{trans}	$f B \times \sin(\gamma)$

includes mixed-polarization states, and optics to analyze the polarization generally follow the telescope entrance. The relevant quantities regarding the effects of spatial resolution for partially polarized light are d, the telescope diameter, and $I \pm P$, where P is any one (or a combination of) circular $[V]$ or linear $[Q, U]$ polarization signals, following the Stokes convention. The optical resolution varies (roughly) linearly with respect to d, meaning that the light which forms the respective Airy disk on a resolution element (a "pixel") is mixed

Table 2 Comparison of "Flux".

Data source	Pixel size (arcsec)	$\sum |B_{los}|dA$ (10^{22} "Mx")	Difference from *Hinode*/SP original (%)
Hinode/SP	0.317	2.80	
Hinode/SP	1.98	2.84	1.2%
MDI	1.98	3.03	8.1%

to an extent determined by aperture size d prior to analysis optics (all other elements in the system being equal). Detected spectra are an intensity-weighted average which is a function of d, meaning that bright contributions will dominate.

3.1. Synthesis and Treatment of Synthetic Spectra

To investigate and demonstrate this effect, we turn first to synthetic data. The approach was briefly described in Leka *et al.* (2009b), and we present it in more detail here. Beginning with a synthetic magnetic model, the effects of different resolution (telescope size) on inferred magnetic field maps are obtained as follows:

- Generate emergent Stokes polarization spectra, $[I, Q, U, V]$ due to the Zeeman effect on a magnetically sensitive photospheric line, assuming a simple Milne-Eddington atmosphere.
- Combine the pure polarized spectra to produce "modulated" spectra $[I \pm P]$, *i.e.*, "observed" mixed-state light.
- Manipulate these spectra as desired, along the lines of:
  - add simulated photon noise by drawing from a Poisson distribution for each particular wavelength, with the expectation value set by the desired "noise level",
  - spatially bin (by summation) the modulated spectra to a desired spatial resolution,
  - average a temporal sequence of modulated spectra from a target location (from a temporal sequence of synthetic maps, as appropriate), and/or
  - apply an instrumental response function.
- Demodulate (combine in linear combination) the manipulated spectra back to pure Stokes $[I, Q, U, V]$.
- Re-invert using the inversion method of choice.

For these tests, spectra were computed using the analytic Unno–Rachkovsky equations applied for the magnetic field vector and velocity at each pixel, and thermodynamic/line parameters typical of the 630.25 nm Fe I spectral line ($g_L = 2.5$, damping $a = 0.4$, Doppler width $\lambda_D = 0.03$ Å, absorption coefficient $\eta_0 = 10$). Generating the Stokes spectra from the model field relied upon the spectra-genesis code which is part of the basic Milne-Eddington least-squares inversion routine "stokesfit.pro" (available from *SolarSoft* distribution[2]). This same inversion was then applied to the resulting Stokes spectra to produce a magnetogram, thus the assumptions underlying the genesis and the inversions for these test data are internally consistent; the goal here is not to test inversion methods *per se*. For the demonstrations here, the manipulation is limited to spatial binning.

3.2. The Magnetic Model

The synthetic magnetic model has a boundary field constructed specifically to include both areas of strong and spatially homogeneous field (reminiscent of sunspot umbrae) and areas

---

[2]http://www.lmsal.com/solarsoft/ssw_whatitis.html.

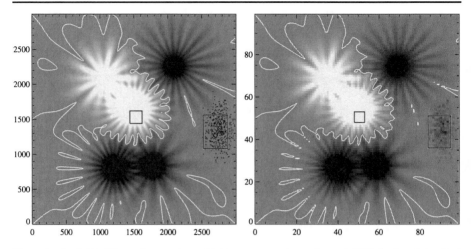

Figure 2 (Left) The "flowers" magnetic model B_z component (saturating at ± 1000 G, $B_z > 0$ is white) at (left:) full resolution, 3000×3000 pixels arbitrarily set to have a $0.03''$ size. Red boxes indicate the sub-regions highlighted in the later analysis, an "umbra" (180×180 pixels) and "plage" (360×480) areas. The smoothed polarity inversion line is shown as a white contour. (Right) Same, but after the spatial rebinning by a factor of 30 to a pixel size of $0.90''$ (using the method of spatially binning the spectra,; see Section 3.1).

with significant fine-scale structure (with features resembling penumbral fibrils and plage area). Nicknamed the "Flowers" model (Figure 2), it is a potential-field construction that fully satisfies Maxwell's equations. It is (generally) resolved on the 3000×3000 computational grid, and a $0.03''$ "pixel size" is assigned arbitrarily; this implies that the magnetic fill fraction is unity for each pixel. This synthetic boundary formed the basis of tests regarding the effects of spatial resolution on ambiguity-resolution algorithms for vector magnetic field data (Leka *et al.*, 2009b). We refer readers to that paper for a detailed description of its construction.

3.3. Signal Mixing in Spatially Averaged Stokes Spectra

The manipulations outlined above are the minimal steps necessary to model the effects of an observing system. Obviously we are completely ignoring the details of a full optical system or spatial smearing due to instrument jitter or atmospheric seeing effects. In addition, in this extremely limited demonstration we are completely ignoring any substantive difference between an imaging system and a slit-spectrograph polarimeter, and we are ignoring photon noise. Of additional note: there are no velocities in this synthetic model, which simplifies the spectral-mixing effects considerably: no asymmetries or additional broadening is introduced to the spectra. In short, the present study uses the simplest possible case.

We perform the spatial binning for a wide range of factors ranging from 2 to 60. We also include a unity bin factor, in order to have a consistent treatment of the spectra/inversion for comparison, rather than comparing to the raw synthetic model; in practice (as discussed in Leka *et al.*, 2009b) only a few pixels of the 9 million in the bin-1 case differ by more than machine precision from the original model field.

The effects of spatial resolution on the detected spectra are demonstrated in Figure 3. Consider two 10×10-pixel portions of the boundary, centered in the "umbra" and in the "plage", respectively. For each, samples from the 100 emergent demodulated Stokes spectra are shown. The emergent spectra for the umbral area are spatially very consistent (Figure 3

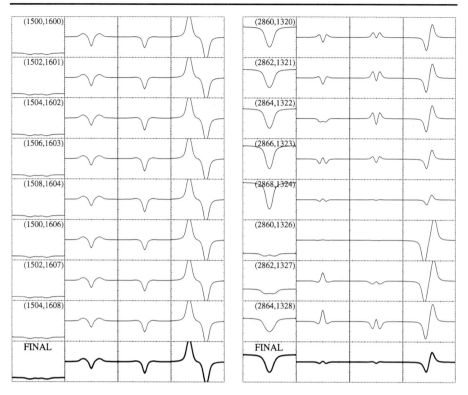

Figure 3 Left column: Eight samples of emergent Stokes [I, Q, U, V] spectra, from a small patch (10×10 pixels) of the original synthetic (fully resolved) data, centered in an "umbra" at [1505, 1605] in Figure 2 (left). Stokes [I, Q, U, V] are plotted left-right with ranges: $I : [0, 1]$, Q, $U : [−0.2, 0.2]$, $V : [−0.5, 0.5]$, the pixel coordinates (of the original model) are also shown. Left, bottom: The resulting "FINAL" [I, Q, U, V] after averaging the 100 underlying emergent polarization spectra, plotted on the same scale. For this case, the resulting average is very similar to any of the sample contributing spectra. Right column: Same as left set, but for a 10×10 pixel area centered on the "plage" area, at [2865, 1325] in Figure 2 (left). In this case, the variability of the underlying spectra (top) leads to an average which differs noticeably from that arising from any single contributing pixel.

left), and the results of averaging the underlying 100 spectra are very similar to any individual contributing emergent spectra. On the contrary, the emergent spectra from the plage area (Figure 3 right) is spatially quite variable. There results a significant difference between the "spatially binned" resulting Stokes spectra and any single emergent spectrum from the underlying area.

Limited resolution causes an intensity-weighted averaging of the emergent Stokes polarization signals. It is often clear (from multiple lobes and extreme asymmetries, see Sanchez Almeida *et al.*, 1996; Sigwarth *et al.*, 1999; Grossmann-Doerth *et al.*, 2000) that the resulting observed spectra are inconsistent with a single magnetic field vector in a simple atmosphere having a linear source function and no additional gradients of any sort within the resolution element (the Milne–Eddington Unno–Rachkovsky assumptions). But sometimes it is not so clear (Sanchez Almeida, 1997). Since the underlying brightness distribution is unknown, untangling the weighting of the contributing spectra is impossible. This quick demonstration clearly cautions that while a strong signal cannot be created from nothing (instrumental

Table 3 Summary and Specifics of Binning Approaches.

Moniker	Algorithm	Code used	Details
"instrument"	Average modulated Stokes spectra	"awnoise.pro"[a] (modified)	"bin-5" implies averaging 5×5 spectra, then inverting
"Post-facto" approaches			
"average"	Simple average $B^i_{new} = N^{-2}_{bin} \sum_{j=1}^{N^2_{bin}} B^i_j$	IDL "rebin" sample=0	Acts on image-plane field components[b] and field strength, fill fraction
"bicubic"	Bicubic Interpolation with $J \times B = 0$	"brebin.pro"[a]	Acts on ambiguity-resolved magnetograms
"sampled"	Simple sampling of image-plane field components[b] and field strength, fill fraction	IDL "congrid" center=1, interpolate=1	If bin is odd: use center point If bin is even: use average of central four points

[a] Available as part of http://www.cora.nwra.com/AMBIGUITY_WORKSHOP/2005/CODES/mgram.tar.

[b] Image-plane field components are defined as $B^i_x = B_{trans} \cos(\phi)$, $B^i_y = B_{trans} \sin(\phi)$, $B^i_z = B_{los}$, and are used to avoid wrap at $\phi = 0, 2\pi$.

and seeing effects aside, as well as any Doppler effects), a small or nonexistent signal can result even when there are strong underlying fields.

3.4. Creating Magnetograms

We now test the effects of the spatial binning of the polarization spectra on the ability of an inversion algorithm to retrieve the underlying structure. The synthetic binned spectra underwent an inversion using "stokesfit.pro"[3] which solves for the magnitude of the field in the instrument-frame B^i_x, B^i_y, B^i_z, and separately the magnetic fill fraction f (see Table 1). The resulting magnetograms were then ambiguity resolved using the minimum-energy code "ME0",[4] described in Leka *et al.* (2009b), Leka, Barnes, and Crouch (2009a). All parameters used for the inversion and ME0 were the same for each resolution (except those that scaled with array size), as it is not the intent to test either the inversion or the ambiguity-resolution algorithms *per se*. What results are vector magnetic field maps that simulate what would be observed through telescopes when solely the aperture size varies.

For comparison we perform three types of *"post facto"* binning on the bin-1 synthetic magnetogram, as summarized in Table 3. The three utilize a simple averaging (referred to as

---

[3] Implementation details: [I, Q, U, V] default relative weighting: $1/[10, 2, 2, 1]$, fill fraction is fit, the initial guess set to the spatially binned parameters from the original model (*i.e.* as close to the solution as possible), "curvefit" specified unless a bad fit returned, in which case "amoeba" and "genetic" algorithms invoked for optimization.

[4] Available at http://www.cora.nwra.com/AMBIG/.

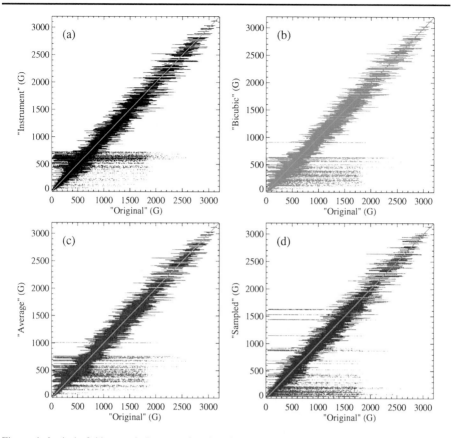

Figure 4 Intrinsic field strength B, comparing the original model magnetogram to the bin-by-30 results, for different binning approaches. (a) Original *vs.* **"instrument"**, (b) original *vs.* "bicubic", (c) original *vs.* "average", and (d) original *vs.* "sampled". For all, the $x = y$ line is also plotted for reference, and on the x-axis ("Original") are plotted all the values represented by the single resulting bin-30 pixel in question, whose value is plotted on the y-axis. Every other point in the binned magnetogram is shown, and every 3rd point of the 900 underlying values is plotted. The colors for these plots will be used consistently below.

"average"), a more sophisticated interpolation method developed by Dr. T. Metcalf specifically for the task of sampling vector magnetograms ("interpolate"), and a sampling approach which performs a minimal amount of averaging (**"sampled"**). We use color here and throughout for reference and clarity as the results of these methods are compared. For each of the "post facto" approaches, the azimuthal ambiguity resolution is an acute-angle method, matched to the results from ME0 for the "instrumental" approach at the same binning factor.

3.5. Comparing the Magnetograms

As seen in Figure 2, spatial rebinning of any sort produces a boxy, somewhat distorted magnetic field map. Quantitatively, however, which of the underlying field's properties are preserved and which are most affected by the change in resolution?

A scatter plot is a good starting place. In Figure 4 the intrinsic field strength B is compared between the original model and the four ways of binning. For all methods, the averaging produces a field that *generally* follows the underlying field distribution; this is reflected

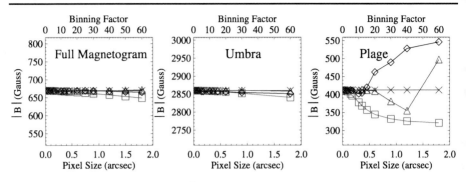

Figure 5 Average intrinsic field strength $\frac{1}{N}\sum B$ as a function of binning factor (top x-axis), for the four binning methods: **"instrument"** ($\Diamond$), "bicubic" ($\square$), "average" ($\times$), and "sampled" ($\triangle$). The three panels show, respectively, the full magnetogram, an "umbral" area and a "plage" area (see Figure 2). For each binning, N varies but the same sub-area of the "Sun" is covered; when non-integer pixel numbers result, that bin factor is omitted. The original model field is sampled at an arbitrarily set $0.03''$, the resulting "pixel sizes" are indicated (bottom x-axis). For these and all similar plots (except where noted), the y-axis ranges are kept consistent between the target areas for direct comparisons. Here, the effects are minimal for the full magnetogram and the "umbra", but have a much larger magnitude and differ between the binning methods in the "plage" area.

in that regression analysis returns a near-unity slope ($\gtrsim 0.98$) for each method. The extremes are lost in what may be termed the "weak-field" areas (up to ≈ 1 kG in the binned case) which are in fact highly structured.

Inversions can sometimes fail to return field strength separately from magnetic fill fraction, especially at low polarization signals. It has been shown that the product of these quantities is significantly more "robust", meaning easier to retrieve reliably (Bommier *et al.*, 2007) (but see also del Toro Iniesta, Orozco Suárez, and Bellot Rubio, 2010). Applying the same regression analysis to the product $f \times B$ indicates that this is not a cure for degraded spatial resolution: slopes and standard deviations which result differ almost imperceptibly, as do the underlying scatter plots, so we do not show them here.

We now examine the inferred magnetic components for the four binning methods ("instrument", "simple", "bicubic", and "sampled") for three target areas ("umbral", "plage", and the full field of view, see Figure 2), as a function of different binning levels. The nature of this comparison is shown in detail in Figure 5. The intrinsic field strength averaged over the (sub)-region in question, $\frac{1}{N}\sum B$ is shown as a function of binning factor for the three target areas. The results for the binning methods are shown for each sub-area. Comparisons following this format are presented for the magnetic fill fraction, the product of the fill fraction and field strength, and the inclination angle distribution (Figure 6). The total unsigned magnetic flux $\Phi = \sum f|B_z|dA$ (Figure 7) is presented, acknowledging the somewhat arbitrary assignment of pixel size. The inferred vertical electric current density $J_z = C\nabla \times fB_h$ was computed for the maps using a finite-difference method that employs a 4-point stencil (Canfield *et al.*, 1993) and C includes all the appropriate physical constants; from this, the total unsigned vertical current $I = \sum |J_z|dA$ is presented (Figure 7) with the same acknowledgement regarding the assigned pixel size as above.

The most significant difference between the plage and umbral areas in the synthetic data is the fact that the former comprises small-scale structure. The umbra area has essentially one magnetic center, whereas the plage area contains a few hundred centers that are highly localized with almost field-free regions separating each center. The different underlying

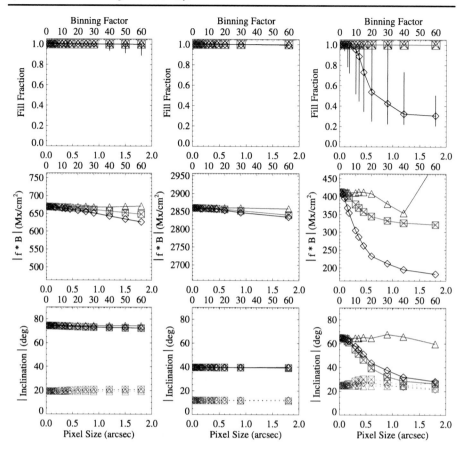

Figure 6 The same format as in Figure 5. Top row: Median (symbols) and 10th, 90th percentiles (displayed as "error bars") of inferred magnetic fill fraction as a function of bin factor. The three "post-facto" approaches consistently return unity since the original model (and bin-1 inversion) have unity fill fraction throughout. Middle row: The average product of the fill fraction and field strength, $\frac{1}{N}\sum fB$ as a function of binning factor. Bottom row: Variation of the average inclination angle with binning factor (thick line-connected curves), $0°$ indicates (unsigned) fields directed along the line of sight, or pure B_{los}, and $90°$ indicates field perpendicular to the line of sight or pure B_{trans} (here, $\gamma = \tan^{-1}(B_{trans}, |B_{los}|)$). Dot-connected curves indicate the standard deviation of the angle distribution.

structure of the field leads to different behavior at different "spatial resolutions", according to the approach.

For field strength (Figure 5), none of the methods show dramatic differences in the umbral area; the same is true for the "full magnetogram". In the plage area, the methods behave quite differently. Simple rebinning shows absolutely no change with bin factor, consistent with its approach of numerically averaging the positive-definite input. The bicubic approach shows a decrease in average field strength, as interpolation increasingly underestimates the strong field strength in the scattered magnetic centers. The sampling follows the simple averaging until approximately bin-20 when it decreases, before abruptly increasing at bin-60. When the bin factor is small, the sum over the subset of sampled points gives a reasonable approximation to the sum over all the (bin-1) points. As bin factor increases, the number of sampled points used to represent the sum decreases, and the result is likely to be increasingly large changes, but with no consistent trend toward increasing or decreasing with bin factor. The instrument binning in the plage area similarly shows minimal effect until approximately bin-10, beyond which the average field strength in the plage area increases. The polarization-free "gaps" between centers begin to be "contaminated" with polarization at higher bin factors, and the resulting average field strength increases, in part because this synthetic plage area is unipolar.

The inversion method separately fits for the field strength and the magnetic fill fraction (Table 1). The synthetic model is fully resolved, so that for bin-1 all pixels return unity fill fraction, and hence all "post-facto" approaches maintain unity fill fraction for all bin factors. When an inversion is performed on spatially averaged spectra, there is almost no effect in the umbral area (Figure 6, top): the median fill fraction remains unity. The situation is very different in the plage area: the non-unity median and wide *range* of fill fraction returned clearly indicate that worsening resolution leads to unresolved structures. The full field of view results reflect a mix of influences from the "resolved" and "unresolved" areas in the field of view.

Whether the underlying structures are resolved or not as indicated by non-unity fill fraction, clearly appears to factor into how worsening spatial resolution will affect the field distribution. The product of fill fraction and field strength (Figure 6) which is arguably a better measure of inversion output, is the same as the field strength for the "post-facto" approaches, but shows a dramatic drop under "instrument" binning. The increase in field strength is more than compensated by a decreasing fill factor, likely as a result of the intensity weighting of the average Stokes spectra.

Other effects of note: the distribution of inclination angle (Figure 6, bottom) with worsening spatial resolution is impacted so as to imply an average orientation closer to the line of sight in the plage than is originally present, for all but the "sample" approach. In other words, with worse spatial resolution the B_{los} begins to dominate over B_{trans}, which might be expected given the lower fractional polarization signal for linear as compared to the circular polarization.

The total magnetic flux (Figure 7) is almost insensitive to bin factor if one uses a post-facto approach, yet plummets with the instrument approach. The sampling approach is slightly variable, again since the value selected will almost randomly hit strong or weak signal as the bin factor increases. Still, the difference is clear: post-facto binning of any kind does not reproduce the effect of spatial resolution.

The total electric current (Figure 7) *increases* with bin factor overall, with a more pronounced effect in unresolved areas than in the unity-fill-fraction umbral region. Recalling that the underlying magnetic model is potential, this somewhat surprising initial increase and the subsequent decrease in plage areas is due to an interplay between the less-smooth map (see Figure 2), and the finite differences used to calculate the vertical current (see the discussion in Leka *et al.*, 2009b); also at play are the influence of the spatial resolution on the relative strength of the horizontal component (as seen through the variation in the inclination angle) and the magnetic fill fraction, which is included when calculating the vertical current density. The bicubic approach, which attempts to include the field structure in the approach, is least affected while the sampling produces the greatest spurious total current. Comparing the results for the umbra and plage sub-area to the full field of view, it is clear that most of the resulting current arises from unresolved areas such as the "penumbra-like" regions that dominate the synthetic model.

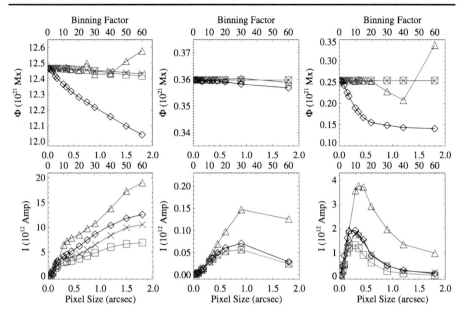

Figure 7 Variation with spatial resolution of parameters often used for characterizing active regions. Top: The total unsigned magnetic flux $\Phi = \sum f |B_z| dA$. Bottom: The total unsigned electric current $I = \sum |J_z| dA$. For these plots, the y-axis ranges vary.

3.5.1. Statistical Tests of Similarity

The question remains how best to characterize the differences in the results at different spatial resolutions. We see from the previous analysis that the resulting magnetograms do differ, but can they adequately describe the underlying field?

We perform Kolmogorov–Smirnoff tests on the distribution of the resulting field parameters to investigate how well a lower-resolution map characterizes the highest-resolution map. The K–S test uses the cumulative probability distribution (CPD) to compare two samples. Two parameters result: "P", the probability of rejecting the null hypothesis, and the "D"-statistic, which measures the maximum difference between the two CPDs. In this case the null hypothesis can be stated, "The two samples arise from the same population", the two samples being, *e.g.*, the map of $B(x, y)$ from the full-resolution data and the map from a binned-resolution magnetogram. It should be remembered that for a given K–S D-statistic, the KS-probability statistic is extremely sensitive to changes in the sample sizes, which is very much the case when the bin factor becomes large.

Comparisons of the CPDs for field strength and vertical current density (Figure 8) confirm that the widest differences imposed at the bin-30 level occur in the plage area. Other parameters (inclination angle, etc.) show similar behavior. The umbral area and the similarity between the CPDs there and the full magnetogram would lead us to believe (correctly, as demonstrated in Figure 6) that for this model, the full magnetogram area is dominated by areas of high fill fraction while still containing areas of unresolved highly structured field.

For the distribution of field strength, the D-statistic (Figure 9) is dominantly zero for the full field of view, and increases only slightly with worse spatial resolution in the umbra. However, it is significantly non-zero for the "plage" area, reflecting that all bin factors show the same behavior seen in detail in Figure 8. The smallest D-statistic in the plage area comes

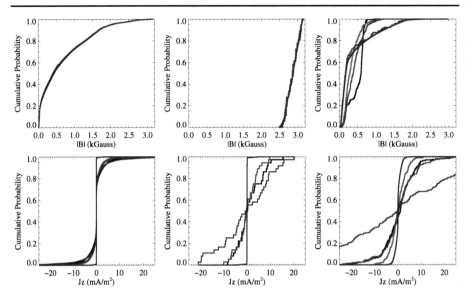

Figure 8 Cumulative probability distributions, comparing that for the full-resolution synthetic map to the bin-30 results, for the three fields of view (entirety, "umbra", and "plage" areas). For each, CPD curves are plotted for the original resolution, the **instrument** method, and the bicubic, average, sampled post-facto approaches. The top row is for the intrinsic field strength B, and the bottom row is for the vertical electric current density J_z.

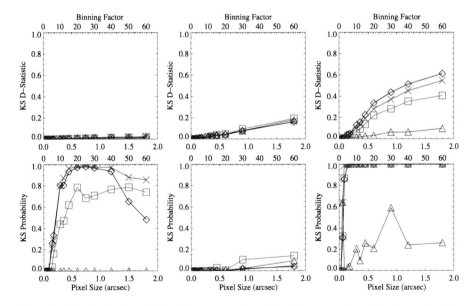

Figure 9 Again for the three fields of view, summaries of the Kolmogorov–Smirnoff tests as a function of binning factor, for the field strength B. The top row shows the D-statistic, and the bottom row shows the probability P that the two samples considered are *different* (see text). Shown are four curves, original resolution *vs.* "instrument" ($\Diamond$), "bicubic" ($\Box$), "average" ($\times$), and "sampled" ($\triangle$) magnetograms.

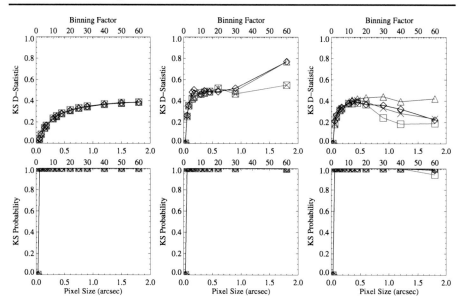

Figure 10 Summaries of the Kolmogorov–Smirnoff tests as a function of binning factor, for the vertical electric current density J_z. Format follows Figure 9.

from the sampling approach; the greatest from the instrumental approach. The probabilities of rejecting the null hypothesis are mixed but generally close to unity for the plage area, with systematically lowest probabilities for sampling, as expected.

For the distribution of J_z, a quantity derived by taking derivatives of the field distribution, the K–S D-statistic (Figure 10) is significantly non-zero for all three sub-areas and all methods at all spatial resolutions. The KS-probability is consistently unity; this bodes ill for the possibility that unresolved magnetograms recover the underlying distribution of field or vertical current.

To summarize these results, in areas such as this model "umbra", the underlying field varies little and the inferred fill fraction is consistent with it being "resolved". It can be argued that through a wide range of spatial resolution, the inferred field distribution represents the underlying field. The situation for highly structured underlying field is very different: areas of low and mixed fill fraction imply that the field is not resolved. It is fairly clear that instrumental effects on the spectra result in a substantively different field distribution, and the implied structures should be treated with much less confidence. And, with all caveats acknowledged due to the use of synthetic data, we find that in general, inferring the distribution of the vertical current is very susceptible to the effects of spatial resolution.

4. Demonstration: Real Data, Revisited

One may always argue that synthetic data constructed to demonstrate a particular effect may not represent observational "truth". Hence, we perform the same exercise using data from the Solar Optical Telescope SpectroPolarimeter aboard the *Hinode* mission (Tsuneta *et al.*, 2008). While the data from this instrument are arguably not the highest-resolution spectro-polarimetric data available, the temporal and spatial consistency coupled with very good resolution in both spatial and spectral dimensions make these data ideal for this purpose.

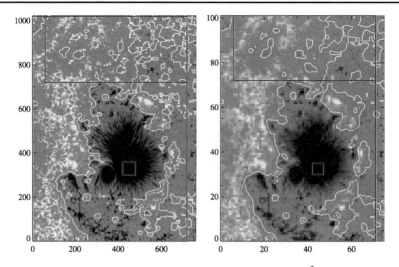

Figure 11 The B_{los} component inferred by *Hinode*/SP, scaled to $\pm 1000\,\mathrm{Mx\,cm}^{-2}$, for NOAA Active Region 10953 observed at 18:35 UT, 30 April 2007. Left: Full-resolution data, with original dimensions 762×1024 and $0.15''$ pixels size. Boxes indicate the sub-regions highlighted in the later analysis, "umbra" (60×60 pixels), "plage" (660×300). In addition, as shown the "full field of view" is slightly trimmed (to 720×1020) to ensure integer divisibility by a range of factors. Right: Same, after "instrument" binning by a factor of 10, to $1.5''$.

We chose the 18:35 UT scan of 30 April 2007 scan of NOAA Active Region (AR) 10953, observed at S09.5, E11.5 ($\mu = 0.98$), which was a "normal" scan that approximately matches scan-steps to the slit width and does not perform any on-board summation. The field of view includes a sunspot and plage area sufficient for this purpose. The pixels are not exactly square, and are not interpolated to be square, but treated as unequal in dimension for all of the analysis; we do, however, use the average of $0.15''$ when referring to general pixel size.

An approach parallel to that described above was used to treat the *Hinode*/SP data, albeit beginning with the fully calibrated Level-1D $[I,\ Q,\ U,\ V]$ Stokes spectra.[5] In this case there is already photon noise present in the data, and the demodulation is performed on-board. In the context of the Poisson-statistics (see Appendix A), the implications are that we cannot exactly model the effects of different apertures. Without the "raw" observed modulated $I \pm P$, $P \in [Q,\ U,\ V]$ spectra and the different contributing realizations of noise, information has already been lost, and manipulating the demodulated pure $[I,\ Q,\ U,\ V]$ spectra is equivalent to reconstructed mixed-polarization states. The manipulated (averaged spatially by summation) spectra will present with lower noise than would actually be the case, but the primary effects of spatial resolution modeling will still be apparent.

The binned spectra were written in the "ASP" format (with a reformatter courtesy B. Lites, HAO/NCAR), and inverted using the HAO/NCAR Milne-Eddington inversion code "sss-inv" (Skumanich and Lites, 1987; Lites and Skumanich, 1990; Lites *et al.*, 1993, with minor modifications for *Hinode*/SP specifics, again courtesy B. Lites, HAO/NCAR).[6] The

---

[5] http://sot.lmsal.com/data/sot/level1d/.

[6] Implementation details: $[I,\ Q,\ U,\ V]$ weighting:$1/[100, 1, 1, 10]$, fill fraction solved, initial guess via "genetic" algorithm optimization, all pixels inverted (no minimum-polarization threshold), "scattered light" profile determined where $\sum |P| < 0.4\%$.

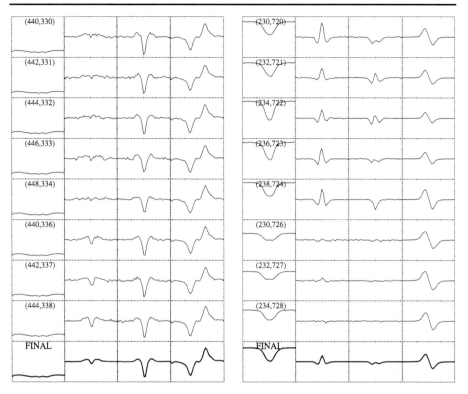

Figure 12 Left column: Eight samples of emergent Stokes [I, Q, U, V] spectra, from a small patch (10×10 pixels) of the full-resolution *Hinode*/SP map, centered in the sunspot umbra at [445, 335] in Figure 11(left). Stokes [I, Q, U, V] are plotted left-right with ranges: $I : [0, 1]$, Q, $U : [-0.2, 0.2]$, $V : [-0.5, 0.5]$, the pixel coordinates (of the original data) are also shown. Left, bottom: The resulting "FINAL" [I, Q, U, V] after averaging the 100 underlying emergent polarization spectra, plotted on the same scale. Right column: Same as left set, but for a 10×10 pixel area centered in the "plage" area, at [235, 725] in Figure 11 (left). For these plage data, the ranges are $I : [0, 1]$, Q, & $U : [-0.1, 0.1]$, $V : [-0.5, 0.5]$.

full-resolution data were subjected to the same reformatting and inversion (without binning) to ensure a consistent comparison. The "ME0" minimum-energy code was used in a consistent manner for ambiguity resolution for all data, and J_z was calculated in exactly the same manner as for the synthetic data. A sample binned magnetogram is shown in Figure 11.

As with the synthetic data, three areas are analyzed: the full field of view, and then separately two areas, one centered on the sunspot umbra and another on a plage area to the north of the sunspot (Figure 11). The latter area was chosen to avoid the emerging filament at the south east edge of the sunspot (Okamoto *et al.*, 2008). The full scan was trimmed slightly and both sub-areas were chosen to be evenly divisible for a number of binning factors.

Samples of the effects of "instrument" binning on emergent Stokes spectra from the *Hinode*/SP data are shown in Figure 12. The umbral sample displays very consistent Stokes spectra, and a final bin-10 result that closely resembles any single constituent-pixel's set of spectra. The noise is nicely reduced in the binned spectra (although somewhat artificially, as described above and in Appendix A). The plage sample demonstrates exactly the effect shown in Figure 3, that the constituent spectra are quite variable, and the resulting binned data reflect an average that does not represent any single underlying pixel.

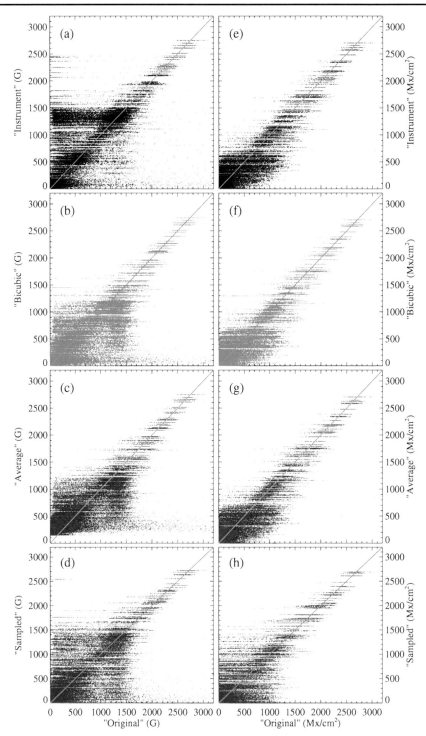

Figure 13 Follows Figure 4 for (a)–(d), except comparing the original *Hinode*/SP data with bin factor 16 results. Figures (e)–(h) follow the same format, but for the product $f \times B$.

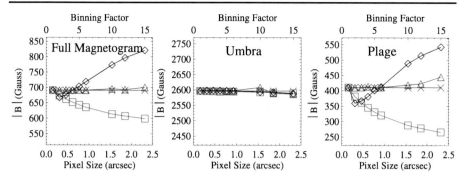

Figure 14 Following Figure 5, the average field strength over the target area, $\frac{1}{N}\sum B$ as a function of binning factor (top x-axis), for the four binning methods (**"instrument"** $\Diamond$, "bicubic" $\Box$, "average" $\times$ and "sampled" $\triangle$), focusing on three areas as indicated: the full magnetogram, an "umbral" area and a "plage" area, as depicted in Figure 11. With the original *Hinode*/SP scan resolution of 0.15″, the resulting pixel sizes are also indicated (bottom x-axis).

Scatter plots of the inverted manipulated spectra demonstrate the general averaging which results with worsening spatial resolution (Figure 13). Of note in the *Hinode*/SP data, compared to the synthetic case (Figure 4), is the much greater spread in the original-resolution field strengths compared to the binned results. This behavior occurs primarily in "weak-field" or weak-polarization areas, where determining the field strength and fill fraction independently is arguably problematic; but that is not the case for all pixels. The product $f \times B$ is also shown; the distributions do change perceptibly (contrary to the synthetic case), with decreased scatter in weak-signal areas. (However, recall that only the "instrument" binning result is an independent inversion.) Primary contenders for the different behavior between B and $f \times B$ here, compared with the synthetic data, include the effects of photon noise and the contention that the original-resolution *Hinode*/SP data are unresolved to begin with.

Changes in the inferred magnetic field distribution in the observational data show similar trends with binning factor as was seen in the synthetic data. Beginning with field strength (Figure 14), the umbral area shows little change, but the plage area is quite sensitive to bin factor and to method used. The full field of view behaves closest to the plage.

The other inferred parameters examined here, the fill fraction, product $f \times B$, and instrument-frame inclination (Figure 15) confirm the general behavior observed in the synthetic-data experiments. The *Hinode*/SP data start with a wide range of inferred fill fraction present, and a median of less than 50% at full resolution for the full field of view (Figure 15). Again, the three post-facto binnings do an averaging or sampling, hence the mean of the fill fraction distribution stays the same although the range of values present decreases with bin factor. The "instrument" binning results in a decreasing mean and tighter range as the spatial resolution degrades, indicating that areas which were resolved become less so.

The product $f \times B$ shows a systematic decrease, on average, with worse spatial resolution – except from the "sampling" approach, which stays relatively constant. The "instrument" approach displays the most variation with resolution change, but the difference between it and the other methods is not as dramatic compared to the experiment with the model data.

The results for field inclination (Figure 15), show a distinct trend of the field becoming more aligned with the line of sight with decreasing spatial resolution, especially in the plage areas. In the umbra, there is effectively no change in the inclination angle distribution. The imperturbability of "sampling" against variations in inclination was seen earlier, as well;

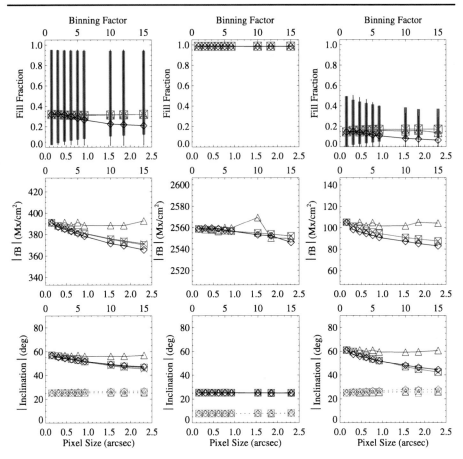

Figure 15 The same format as in Figure 14. Top row: Median (symbols) and 10th, 90th percentiles (displayed as "error bars") of inferred magnetic fill fraction as a function of bin factor. The three "post-facto" approaches consistently return the same fill fraction as the original observations, as expected. Middle row: The average product of the fill fraction and field strength, $\frac{1}{N} \sum fB$ as a function of binning factor. Bottom row: Variation of the average inclination angle with binning factor (thick line-connected curves), 0° indicates (unsigned) fields directed along the line of sight, or pure B_{los}, and 90° indicates field perpendicular to the line of sight or pure B_{trans} (here, $\gamma = \tan^{-1}(B_{trans}, |B_{los}|)$). Dot-connected curves indicate the standard deviation of the angle distribution.

again, the sampling should represent the underlying field distribution (until the super-pixels are themselves large and the resulting number of binned pixels available is small), since it samples rather than averages. We present the image-plane inclination angle from the line of sight – closely related to the direct observables, but related to the physical inclination of the field to the local normal by way of the observing angle. Since $\mu = 0.98$ for these data, the difference between image-plane and the heliographic-plane inclination from the local vertical direction is minimal.

The total unsigned magnetic flux, $\Phi = \sum f |B_z| dA$ behaves essentially the same in the umbral areas of both synthetic and *Hinode* data, varying little with resolution (except when there are arguably very few points within that area of interest, see Figure 16). And again, the full field-of-view behavior is dictated by what kind of structure dominates at highest resolution. We see that the "instrument" spectral binning and subsequent inversion, which

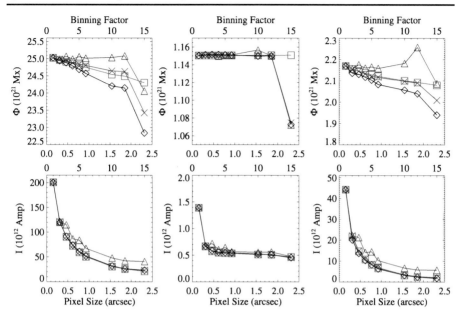

Figure 16 Comparison of parameters often used for characterizing active regions. Top: Variation of the total unsigned magnetic flux $\Phi = \sum f |B_z| dA$ Bottom: Variation of the total unsigned electric current $I = \sum |J_z| dA$. For these plots, the y-axis is allowed to vary.

is designed to mimic decreasing telescope size, produces more of an effect than the "post-facto" binning approaches.

The total vertical electric current is often used to parametrize an active region's stored magnetic energy (Leka and Barnes, 2003 and references therein). Could this characterization differ as a function of spatial resolution? In the *Hinode*/SP data, for all fields of view, there is a smooth decrease of total current with decreasing spatial resolution. In addition, all binning methods appear to act identically in this case. The behavior of the *Hinode*/SP data most resembles the synthetic "plage" beyond bin factor 10. That is, the observational data, even at 0.15″, most closely resembles the area filled with unresolved multiple small-scale magnetic centers.

Overall, the plage area observed with *Hinode*/SP produces the most variations due to rebinning or degraded spatial resolution. The umbral area is least sensitive. The sampling typically provides the most consistent answer, but is also susceptible to the particular point sampled. The effect of changing the instrumental resolution more closely follows the results of the post-facto approaches as compared to the trends in the simulation data. Assuming that the behavior of the full magnetogram is characterized by the relative fraction of "resolved" or near-unity fill fraction pixels within the field of view, it is clear that the *Hinode*/SP data are dominated by non-unity fill fraction pixels and unresolved field structure, even at the highest resolution.

The Kolmogorov–Smirnoff tests confirm statistically what is described above. The cumulative probability curves for field strength (Figure 17), comparing the bin-10 results to the original resolution for both "instrument" and post-facto binnings indicate distinct differences in the full field of view which is reminiscent of the behavior in the plage area. The umbral field strength CPD looks almost identical to the umbral CPD for the synthetic data (Figure 8). The distribution of the vertical current density (Figure 17) shows an almost

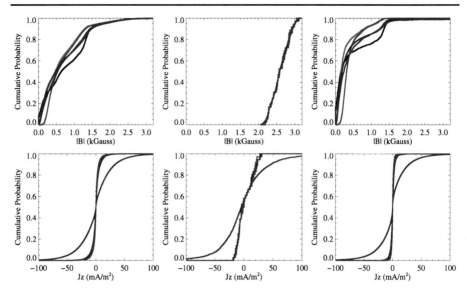

Figure 17 Cumulative probability distributions, for the full-resolution data and the bin-10 results, for the three fields of view (entirety, "umbra", and "plage" areas). For each, CPD curves are plotted for: original resolution, **instrument binning**, bicubic, average and sampled approaches. The top row is for the intrinsic field strength B, and the bottom row is for the vertical electric current density J_z.

exactly opposite behavior than was observed in the synthetic data, in that the original resolution indicates the presence of inferred vertical current which has decreased in magnitude significantly at bin factor 10.

The K–S statistics for field strength are more consistent across bin factors (Figure 18) than in the synthetic data: the D-statistic is slightly elevated but only varies dramatically with the simple binning. The K–S probability is unity for the plage area and the full field of view for all bin factors, indicating that the samples are not drawn from the same population. On the contrary, it can be argued that areas with consistent unity fill fraction statistically sample the same population as the underlying field.

On the other hand, the vertical current density is affected at all spatial resolutions (Figure 19). From a statistical point of view the results from lower resolution data do not represent the underlying distribution of the highest spatial resolution, even in the unity-fill-fraction umbral area. One may simply conclude that the actual distribution of vertical current in the solar photosphere is unknown and unknowable without absolutely full resolution everywhere in question.

5. Summary and Conclusions

We outline a manner by which to manipulate Stokes polarization spectra in order to mimic the effects of instrumental spatial resolution to the simplest order. Through the use of a synthetic magnetic field construct that is both fully resolved and contains small-scale structures, we apply this method to a range of degradations. We find (not surprisingly) that it is the highly structured areas which are most sensitive to the effects of instrumental optical spatial resolution.

The analysis indicates (also not surprisingly) that even the *Hinode*/SP "normal scan" spectro-polarimetric data at 0.15″ spatial sampling are unresolved. Recalling this, plus the

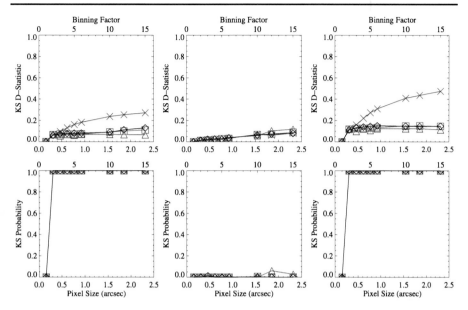

Figure 18 Summaries of the Kolmogorov–Smirnoff tests as a function of binning factor, for the intrinsic field strength B over the three fields of view. The top row shows the D-statistic, and the bottom row shows the probability P that the two samples are *different* (see text). Shown are curves for the original resolution *vs.* the (**"instrument"** ◊, "bicubic" □, "average" × and "sampled" △) vertical current distributions.

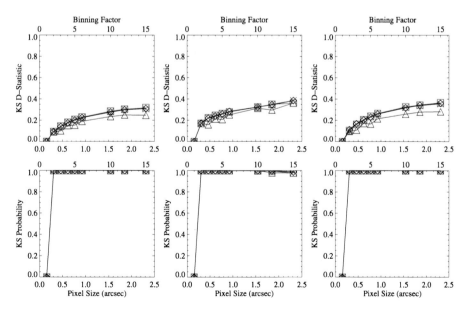

Figure 19 Summaries of the Kolmogorov–Smirnoff tests as a function of binning factor, for the vertical electric current density J_z, following Figure 18.

fact that we could only bin up to a factor of 16 before completely decimating the number of pixels needed for analysis, the patterns shown by field parameters with degrading resolution are remarkably similar to those found using the synthetic data. We thus confirm the appropriateness of the findings from these synthetic-data as valid for helping interpret the observational data.

Statistical tests confirm that whether by instrumental spectral mixing or post-facto methods, worsening spatial resolution results in a map of the vector field which does not reproduce the underlying magnetic structure, except in select areas where the returned magnetic filling factor in the binned data is still unity. Where the returned fill factor is less than unity, worsening spatial resolution leads to an average image-plane inclination angle more aligned to the line of sight, an increasing average field strength which couples with the decreasing average fill fraction to present a decreasing total magnetic flux. The behavior of further-derived parameters that rely on spatial derivatives is less straightforward, but may impart a non-zero current density and inferred "twist" where there in fact are none. The pessimistic interpretation of these results is that without the highest spatial resolution, the underlying field is unrecoverable. The optimistic interpretation is that by making use of the inferred magnetic fill fraction for inverted spectro-polarimetric data, it is possible to tell where these effects will be most dramatic, and where they will be least impactful.

The influence of spatial resolution on the instrument-plane inclination angle implies that the impacts on physically interpreted variables in the (coordinate-transformed) heliographic plane will vary with observing angle. This also has implications for our understanding of the large-scale "weak-field" areas from instruments of limited spatial resolution: in this context, the assumption that the photospheric field is dominantly radial (Wang and Sheeley, 1992; Arge et al., 2002) must be re-examined.[7]

Details and caveats to the above statements are important to mention. There is no model of instrumental scattered light applied to (or subsequently corrected for) the synthetic data; in parallel, the *Hinode*/SP data are inverted using a common but simple treatment of computing a scattered light profile, rather than a more sophisticated local approach which has been demonstrated to better recover low-signal areas (Orozco Suárez et al., 2007). While the details will differ had we used the latter, the approach taken here is consistent, and hence still illustrative. Effects as drastic as shown here are generated in the synthetic data without Doppler velocities or field gradients along the line of sight, whereas both are expected for observational data. Yet in the "simple is OK" defense, key behavior patterns are seen clearly in the *Hinode*/SP data.

We also ask how well instrumental resolution can be represented by "post-facto" manipulation of the vector-field map. Tests of three different methods show that, again, in highly structured underlying areas, these methods result in very different outcomes than expected from differences in aperture size. Simply put, there are only special cases where "binning down" a magnetogram will adequately mimic the differences between different instrumental spatial resolutions, and generally the "instrument" binning results in the largest differences from the underlying field.

This exercise of comparing the results when one simulates "worse spatial resolution" by different means is illuminating, and demonstrates that method matters according to the goal of the study in question. Three basic categories are: comparisons/calibrations between instruments, utilizing data from different instruments as part of an analysis for which data from a single instrument falls short (due to limited field of view, capability, availability, etc.), and interpreting numerical simulation results in the context of observations.

---

[7] As are the results that they may be predominantly horizontal, see Borrero and Kobel (2011).

Regarding the first category, we note that while a few instrument-comparison studies perform spatial averaging on the polarization signals for comparison (Wang *et al.*, 1992; Labonte, Mickey, and Leka, 1999),[8] the majority such studies published thus far use some form of "post-facto" averaging and binning applied to the magnetogram from the higher-resolution instrument (Berger and Lites, 2002, 2003; Tran *et al.*, 2005; Demidov *et al.*, 2008; Wang *et al.*, 2009a, 2009b). It is clear that "instrument" binning should be the preferred method, since all post-facto approaches result in a different (and typically smaller) variation with binning factor than expected from optical resolution.

An addendum to this category is using synthetic data for tests of algorithms through "hare & hound" exercises, where the evaluation depends crucially that the synthetic data mimic the behavior of those real data eventually slated for analysis. As such, including the gross effects of the instrument or observing method chosen (Leka *et al.*, 2009b; Orozco Suárez *et al.*, 2007) is needed in order to not arrive at incorrect conclusions.

In the second category, if the goal is to preserve the underlying character of the vector magnetic field *and* the region in question has a high average filling fraction, then post-facto binning can be employed with some confidence. However, as was shown with the vertical current density, while the magnetic field distribution and character may be preserved, quantities that are derived from the field must be viewed with less confidence. This is a very restrictive set of caveats, but the most well-defended position according to this study.

The third category acknowledges the great strides in simulations of solar magnetic structure, and the approach of validating them quantitatively using comparisons to observed structures (Leka and Steiner, 2001; Abbett, 2007; Orozco Suárez *et al.*, 2007; Sheminova, 2009). It is insufficient to rebin or apply a blurring function directly to a simulation's well-resolved output for comparisons to the solar observations. We reiterate that, due to these results, at the very least a simple modeling and manipulation of emergent spectra is required for even qualitative comparisons between simulations and observations of the magnetic field distribution.

In this context, we come back to Table 2 (see Appendix B, Table 4 and Figure 20). The minimal impact of the post-facto "congrid" approach on the *Hinode*/SP fast-scan map "Flux" $= \sum |B_{los}|$ result is consistent with what we have shown here. The MDI Level 1.8.1 data used in Section 2 and in De Rosa *et al.* (2009) present a systematic offset from the Level 1.8.2 calibration (which became available December 2008, and decreased the B_{los} magnitudes by $\approx$ 8–9% in the location of AR 10953[9]). When variations in field of view, calibration, and especially spatial resolution are accounted for according to the findings of this paper (details can be found in Appendix B), there still exists an offset between the results from MDI and *Hinode*/SP that is larger than the quoted uncertainties, but may still be attributable to remaining differences in the lines' formation heights and inversion methods.

Finally, from this investigation, it is still unclear what the solar magnetic field structure actually *is*, especially for areas with fine-scale structure. This is not a new concept (Sánchez Almeida and Lites, 2000), but reinforced here through a simple, yet thorough demonstration. We show that our ignorance is especially true for quantities derived from the vector-field maps which rely on spatial derivatives (Parker, 1996; Leka *et al.*, 2009b). Are vector magnetic field maps useless? Definitely not! Comparisons between data of active regions obtained with consistent instrumentation and spatial resolution do detect differences amongst

---

[8]Labonte, Mickey, and Leka (1999) performed a near-simultaneous comparison between the IVM and the ASP, contrary to the note in Berger and Lites (2002), Section 1.1.

[9]See http://soi.stanford.edu/magnetic/Lev1.8/ for details.

the structures that must, somehow, be related to the inherent magnetic structure, especially as manifest in the release of stored magnetic energy (see, *e.g.*, Leka and Barnes, 2007 and references therein). But in the context of measuring and interpreting the state and behavior of the solar plasma, conclusions that are drawn must do so in the context of the limitations of the data employed.

Acknowledgements   KDL first acknowledges Dr. Richard C. Canfield, who introduced her to spectro-polarimetry and the interpretation of vector magnetic field maps (at 6″ resolution!). We also appreciate the supportive and helpful comments from the referee. This work was made possible by the models and instruments developed under the following respective funding sources: NASA contracts NNH05CC75C, NNH09CE60C and NNH09CF22C, the NWRA subcontract from the Smithsonian Astrophysical Observatory under NASA NNM07AB07C, and the NWRA subcontract from Stanford University NASA Grant NAS5-02139 for SDO/HMI commissioning and pipeline code implementation. We thank Dr. Bruce Lites at NCAR/HAO for reformatter code and updates to the HAO inversion code. We also sincerely thank Mr. Eric Wagner for understanding "scientist code" and making things actually run. *Hinode* is a Japanese mission developed and launched by ISAS/JAXA, collaborating with NAOJ as a domestic partner, NASA and STFC (UK) as international partners. Scientific operation of the *Hinode* mission is conducted by the *Hinode* science team organized at ISAS/JAXA, consisting of scientists from institutes in the partner countries. Support for the post-launch operation is provided by JAXA and NAOJ (Japan), STFC (U.K.), NASA, ESA, and NSC (Norway). MDI data are provided by the SOHO/MDI consortium. SOHO is a project of international cooperation between ESA and NASA.

Appendix A: Constructing Representative Instrument-Binned Spectra When Only Demodulated Spectra Are Available

For an instrument like *Hinode*/SP, the demodulation from six states is performed on-board the spacecraft, so only the four demodulated states are available. Since the demodulated states do not contain all the information of the original states, we discuss here the impact of this loss of information on the noise level of the reconstructed states.

Assuming that each of the six polarization states actually observed at a given wavelength, $(I \pm P)(\lambda)$, is drawn from a Poisson distribution, the expectation value of each distribution is given by $\langle I \pm P \rangle \equiv p_{\pm}^{\lambda}$, where P can be any of Q, U or V. Since each of these is a Poisson distribution, the variance of each is equal to the expectation value.

The demodulated states actual available are given by

$$P(\lambda) \equiv \left[(I + P)(\lambda) - (I - P)(\lambda)\right]/2 \tag{1}$$

$$I(\lambda) \equiv \left[(I + Q)(\lambda) + (I - Q)(\lambda) + (I + U)(\lambda) + (I - U)(\lambda)\right.$$

$$\left. + (I + V)(\lambda) + (I - V)(\lambda)\right]/6. \tag{2}$$

(Henceforth, the wavelength dependence is assumed for clarity.) Working specifically with $I \pm Q$ as an example, since each modulated state will have similar behavior, the reconstructed modulated states are

$$(I \pm Q)_{\mathrm{R}} = I \pm Q$$

$$= \frac{1}{6}\left[(I + Q) + (I - Q) + (I + U) + (I - U)\right.$$

$$\left. + (I + V) + (I - V)\right] \pm \frac{1}{2}\left[(I + Q) - (I - Q)\right]$$

$$= \frac{2}{3}(I \pm Q) - \frac{1}{3}(I \mp Q)$$

$$+ \frac{1}{6}\left[(I + U) + (I - U) + (I + V) + (I - V)\right] \tag{3}$$

which has an expectation value of

$$\langle (I \pm Q)_{\mathrm{R}}(\lambda) \rangle = \frac{2q_{\pm}^{\lambda}}{3} - \frac{q_{\mp}^{\lambda}}{3} + \frac{u_{+}^{\lambda} + u_{-}^{\lambda} + v_{+}^{\lambda} + v_{-}^{\lambda}}{6}, \tag{4}$$

and a variance of

$$\mathrm{var}(I \pm Q)_{\mathrm{R}}(\lambda) = \frac{4q_{\pm}^{\lambda}}{9} + \frac{q_{\mp}^{\lambda}}{9} + \frac{u_{+}^{\lambda} + u_{-}^{\lambda} + v_{+}^{\lambda} + v_{-}^{\lambda}}{36}, \tag{5}$$

whereas the expectation value and the variance of the actual state is simply $q_{\pm}^{\lambda}$. In the continuum (or anywhere the polarization is low), this reduces to

$$\mathrm{var}(I \pm Q)_{\mathrm{R}}^{\mathrm{c}} = \frac{4q_{\pm}^{\mathrm{c}}}{9} + \frac{q_{\mp}^{\mathrm{c}}}{9} + \frac{u_{+}^{\mathrm{c}} + u_{-}^{\mathrm{c}} + v_{+}^{\mathrm{c}} + v_{-}^{\mathrm{c}}}{36}$$

$$\approx \frac{2}{3}p^{\mathrm{c}}. \tag{6}$$

Thus the variance in the reconstructed modulated states, at least in areas of weak polarization, is smaller than the variance in the original states. Further, since each reconstructed state is the sum of six Poisson variables, rather than being a single Poisson variable (at a given wavelength), the distribution of the noise will also differ.

Appendix B: Comparing MDI and *Hinode*/SP Line-of-Sight "Flux"

As presented in this manuscript, instruments with different resolutions will provide quantitatively different descriptions of the solar magnetic field. We began the study with a provocative "why are these the same, and why are those different?" example. In detail, of course, there is more to this than simply the spatial resolution of two different instruments. The MDI data used for De Rosa *et al.* (2009) were from the level 1.8.1 calibration, the *Hinode*/SP data were provided by B.W. Lites, with ostensibly the same inversion that was used here for the "instrument"-binning exercise (although probably with slightly different implementation), but which also included a remapping to square pixels using an unknown method. Not only the spatial sampling but the field of view differs between the *Hinode*/SP scans of 18:35 and 22:30 UT, as one can see by closely examining Figures 1 and 11.

Here we demonstrate just how sensitive comparisons can be to the details of calibration, inversion, and very slight variations in the physical area sampled. Table 4 summarizes differences in the data sources and processing to obtain maps of AR 10953 on 30 April 2007, and Figure 20 shows the variation in the inferred $\Phi_{\mathrm{los}} = \sum |B_{\mathrm{los}}| dA$ for each. A full propagation of uncertainties was performed, for MDI following Hagenaar (2001), for *Hinode*/SP data using the uncertainties returned from inversions and propagated for B_{los}. For most points, the uncertainty is smaller than the plotting symbol.

The entries are combined into three rough groups. The first is based on the 18:35 UT *Hinode*/SP "normal" scan used for most of this paper, and the total $\sum |B_{\mathrm{los}}| dA$ based on it is deemed the reference. Entries include the results from post-facto "sampling" to match

Figure 20 The total unsigned "flux" $\Phi_{los} = \sum |B_{los}| dA$ for NOAA AR 10953 on 30 April 2007 from various sources and methods of spatial resolution modeling. See Table 4 and text for descriptions of tags and the three groups indicated by vertical lines. Formal error bars are included for each point.

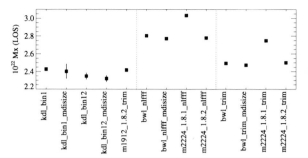

Table 4 AR 10953, 30 April 2007, Total $\sum |B_{los}| dA$ Details.

Label	Data source/Time	Details/Area	Difference (%)
kdl_bin1	*Hinode*/SP 18:35 "normal" scan	"instrument" bin-1	–
kdl_bin1_mdisize	" "	"kdl_bin1"+congrid → 1.98″	−1
kdl_bin12	" "	"instrument" bin-12	−3
kdl_bin12_mdisize	" "	"kdl_bin12"+congrid → 1.98″	−4
m1912_1.8.2_trim	MDI 19:12 UT[a] Level 1.8.2	[387:446,429:525][b]	−0.4
bwl_nlfff	*Hinode*/SP 22:30 "fast" map	Inversion by B.W. Lites for De Rosa *et al.* (2009). Remapped to square pixels	+15
bwl_nlfff_mdisize	" "	"bwl_nlfff"+congrid→ 1.98″	+14
m2224_1.8.1_nlfff	MDI 22:24 UT[c] Level 1.8.1	[385:460,429:509]	+25
m2224_1.8.2_nlfff	Level 1.8.2	" "	+14
bwl_trim	*Hinode*/SP 22:30 "fast" map	Trimmed in x-dir to match *Hinode*/SP 18:35	+3
bwl_trim_mdisize	" "	"bwl_trim"+congrid→ 1.98″	+2
m2224_1.8.1_trim	MDI 22:24 UT[c] Level 1.8.1	[400:461,428:509]	+13
m2224_1.8.2_trim	Level 1.8.2	" "	+3

[a]fd_M_96m_01d.5232.0012.fits.
[b]Indexing starts at 0.
[c]fd_M_96m_01d.5232.0014.fits.

the MDI resolution, and an "instrument" bin-12 to get close to the MDI resolution, with an additional sampling from that to match it exactly as indicated. These are compared to the MDI level 1.8.2 dataset closest in time.

The second and third groups are based on the 22:30 *Hinode*/SP "fast" scan (see Section 2), a post-facto "sampled" map based on it, and comparisons to the closest-time MDI level-1.8.1 and level-1.8.2 data. The difference between these two groups is whether the full 22:30 *Hinode*/SP is used or whether all are trimmed to match the (slightly) smaller field of view of the 18:35 UT *Hinode*/SP scan. The difference is not so slight.

Clearly, the binning approaches behave as described in the text, however those effects are insignificant as compared to even small discrepancies in the field of view and calibration. And evolution: comparing datasets which are as consistent as possible but separated by time, we see the active region increasing its total magnetic signal during this period (see Okamoto *et al.*, 2008).

The answer to the small puzzle presented in Section 2 is that in fact the level-1.8.1 MDI calibration produced systematically larger Φ_{los} results than could otherwise be explained by spatial resolution issues; this is mostly accounted for by the recalibrated MDI level-1.8.2 data, as these examples show. And as we have demonstrated, post-facto manipulation of a magnetogram as in Section 2 does not generally reproduce the differences in instrumental spatial resolution. There exists still a small offset such that the level-1.8.2 data return a $\Phi_{\text{los}} = \sum |B_{\text{los}}| dA$ greater by a few percent than expected from the quoted uncertainties when the best possible match is compared (Table 4, "kdl_bin12" and "kdl_bin12_mdisize" vs. "m2224_1.8.2_trim"). We acknowledge that this is a single example, and invoke spectral-line properties and inversion method differences as probable contributors.

References

Abbett, W.P.: 2007, *Astrophys. J.* **665**, 1469. doi:10.1086/519788.

Arge, C.N., Hildner, E., Pizzo, V.J., Harvey, J.W.: 2002, *J. Geophys. Res.* **107**, 1319. doi:10.1029/2001JA000503.

Bao, S.D., Pevtsov, A.A., Wang, T.J., Zhang, H.Q.: 2000, *Solar Phys.* **195**, 75. doi:10.1023/A:1005244700895

Berger, T.E., Lites, B.W.: 2002, *Solar Phys.* **208**, 181. doi:10.1023/A:1020537923728

Berger, T.E., Lites, B.W.: 2003, *Solar Phys.* **213**, 213. doi:10.1023/A:1023953716633

Bommier, V., Landi Degl'Innocenti, E., Landolfi, M., Molodij, G.: 2007, *Astron. Astrophys.* **464**, 323. doi:10.1051/0004-6361:20054576.

Borrero, J.M., Kobel, P.: 2011, *Astron. Astrophys.* **527**, A29. doi:10.1051/0004-6361/201015634.

Canfield, R.C., de La Beaujardiere, J.-F., Fan, Y., Leka, K.D., McClymont, A.N., Metcalf, T.R., Mickey, D.L., Wuelser, J.-P., Lites, B.W.: 1993, *Astrophys. J.* **411**, 362. doi:10.1086/172836.

De Rosa, M.L., Schrijver, C.J., Barnes, G., Leka, K.D., Lites, B.W., Aschwanden, M.J., *et al.*: 2009, *Astrophys. J.* **696**, 1780. doi:10.1088/0004-637X/696/2/1780.

del Toro Iniesta, J.C., Orozco Suárez, D., Bellot Rubio, L.R.: 2010, *Astrophys. J.* **711**, 312. doi:10.1088/0004-637X/711/1/312.

Demidov, M.L., Balthasar, H.: 2009, *Solar Phys.* **260**, 261. doi:10.1007/s11207-009-9443-5.

Demidov, M.L., Golubeva, E.M., Balthasar, H., Staude, J., Grigoryev, V.M.: 2008, *Solar Phys.* **250**, 279. doi:10.1007/s11207-008-9225-5.

Grossmann-Doerth, U., Schüssler, M., Sigwarth, M., Steiner, O.: 2000, *Astron. Astrophys.* **357**, 351.

Hagenaar, H.J.: 2001, *Astrophys. J.* **555**, 448. doi:10.1086/321448.

Kosugi, T., Matsuzaki, K., Sakao, T., Shimizu, T., Sone, Y., Tachikawa, S., *et al.*: 2007, *Solar Phys.* **243**, 3. doi:10.1007/s11207-007-9014-6.

Labonte, B., Mickey, D.L., Leka, K.D.: 1999, *Solar Phys.* **189**, 1. doi:10.1023/A:1005202503425

Leka, K.D.: 1999, *Solar Phys.* **188**, 21. doi:10.1023/A:1005130630873

Leka, K.D., Barnes, G.: 2003, *Astrophys. J.* **595**, 1277. doi:10.1086/377511.

Leka, K.D., Barnes, G.: 2007, *Astrophys. J.* **656**, 1173. doi:10.1086/510282.

Leka, K.D., Steiner, O.: 2001, *Astrophys. J.* **552**, 354. doi:10.1086/320445.

Leka, K.D., Barnes, G., Crouch, A.: 2009a, In: Lites, B., Cheung, M., Magara, T., Mariska, J., Reeves, K. (eds.) *The Second Hinode Science Meeting: Beyond Discovery-Toward Understanding*, ASP Conf. Ser. **415**, 365.

Leka, K.D., Barnes, G., Crouch, A.D., Metcalf, T.R., Gary, G.A., Jing, J., Liu, Y.: 2009b, *Solar Phys.* **260**, 83. doi:10.1007/s11207-009-9440-8.

Lites, B.W., Skumanich, A.: 1990, *Astrophys. J.* **348**, 747. doi:10.1086/168284.

Lites, B.W., Elmore, D.F., Seagraves, P., Skumanich, A.P.: 1993, *Astrophys. J.* **418**, 928. doi:10.1086/173450.

Moon, Y., Kim, Y., Park, Y., Ichimoto, K., Sakurai, T., Chae, J., *et al.*: 2007, *Publ. Astron. Soc. Japan* **59**, 625.

Okamoto, T.J., Tsuneta, S., Lites, B.W., Kubo, M., Yokoyama, T., Berger, T.E., *et al.*: 2008, *Astrophys. J. Lett.* **673**, 215. doi:10.1086/528792.

Orozco Suárez, D., Bellot Rubio, L.R., Del Toro Iniesta, J.C., Tsuneta, S., Lites, B., Ichimoto, K., *et al.*: 2007, *Publ. Astron. Soc. Japan* **59**, 837.

Parker, E.N.: 1996, *Astrophys. J.* **471**, 485. doi:10.1086/177983.

Sanchez Almeida, J.: 1997, *Astrophys. J.* **491**, 993. doi:10.1086/304999.

Sánchez Almeida, J., Lites, B.W.: 2000, *Astrophys. J.* **532**, 1215. doi:10.1086/308603.

Sanchez Almeida, J., Landi Degl'Innocenti, E., Martinez Pillet, V., Lites, B.W.: 1996, *Astrophys. J.* **466**, 537. doi:10.1086/177530.

Scherrer, P.H., Bogart, R.S., Bush, R.I., Hoeksema, J.T., Kosovichev, A.G., Schou, J., *et al.*: 1995, *Solar Phys.* **162**, 129. doi:10.1007/BF00733429.

Sheminova, V.A.: 2009, *Solar Phys.* **254**, 29. doi:10.1007/s11207-008-9286-5.

Sigwarth, M., Balasubramaniam, K.S., Knölker, M., Schmidt, W.: 1999, *Astron. Astrophys.* **349**, 941.

Skumanich, A., Lites, B.W.: 1987, *Astrophys. J.* **322**, 473. doi:10.1086/165743.

Tran, T., Bertello, L., Ulrich, R.K., Evans, S.: 2005, *Astrophys. J. Suppl.* **156**, 295. doi:10.1086/426713.

Tsuneta, S., Ichimoto, K., Katsukawa, Y., Nagata, S., Otsubo, M., Shimizu, T., *et al.*: 2008, *Solar Phys.* **249**, 167. doi:10.1007/s11207-008-9174-z.

Ulrich, R.K., Evans, S., Boyden, J.E., Webster, L.: 2002, *Astrophys. J. Suppl.* **139**, 259. doi:10.1086/337948.

Ulrich, R.K., Bertello, L., Boyden, J.E., Webster, L.: 2009, *Solar Phys.* **255**, 53. doi:10.1007/s11207-008-9302-9.

Varsik, J.R.: 1995, *Solar Phys.* **161**, 207. doi:10.1007/BF00732067.

Wang, D., Zhang, M., Li, H., Zhang, H.: 2009a, *Sci. China G, Phys. Mech. Astron.* **52**, 1707. doi:10.1007/s11433-009-0249-0.

Wang, D., Zhang, M., Li, H., Zhang, H.Q.: 2009b, *Solar Phys.* **260**, 233. doi:10.1007/s11207-009-9441-7.

Wang, H., Varsik, J., Zirin, H., Canfield, R.C., Leka, K.D., Wang, J.: 1992, *Solar Phys.* **142**, 11. doi:10.1007/BF00156630.

Wang, Y., Sheeley, J.N.R.: 1992, *Astrophys. J.* **392**, 310. doi:10.1086/171430.

Zhang, H., Labonte, B., Li, J., Sakurai, T.: 2003, *Solar Phys.* **213**, 87. doi:10.1023/A:1023246421309

Solar Phys (2012) 277:119–130
DOI 10.1007/s11207-011-9764-z

Magnetic Connectivity Between Active Regions 10987, 10988, and 10989 by Means of Nonlinear Force-Free Field Extrapolation

Tilaye Tadesse · T. Wiegelmann · B. Inhester · A. Pevtsov

Received: 17 December 2010 / Accepted: 28 March 2011 / Published online: 12 May 2011
© Springer Science+Business Media B.V. 2011

Abstract Extrapolation codes for modelling the magnetic field in the corona in Cartesian geometry do not take the curvature of the Sun's surface into account and can only be applied to relatively small areas, *e.g.*, a single active region. We apply a method for nonlinear force-free coronal magnetic field modelling of photospheric vector magnetograms in spherical geometry which allows us to study the connectivity between multi-active regions. We use Vector Spectromagnetograph (VSM) data from the Synoptic Optical Long-term Investigations of the Sun (SOLIS) survey to model the coronal magnetic field, where we study three neighbouring magnetically connected active regions (ARs 10987, 10988, 10989) observed on 28, 29, and 30 March 2008, respectively. We compare the magnetic field topologies and the magnetic energy densities and study the connectivities between the active regions. We have studied the time evolution of the magnetic field over the period of three days and found no major changes in topologies, as there was no major eruption event. From this study we have concluded that active regions are much more connected magnetically than the electric current.

Keywords Corona · Magnetic fields · Photosphere

Solar Flare Magnetic Fields and Plasmas
Guest Editors: Y. Fan and G.H. Fisher.

T. Tadesse (✉) · T. Wiegelmann · B. Inhester
Max Planck Institut für Sonnensystemforschung, Max-Planck Str. 2, 37191 Katlenburg-Lindau,
Germany
e-mail: tilaye.tadesse@gmail.com

T. Wiegelmann
e-mail: wiegelmann@mps.mpg.de

B. Inhester
e-mail: inhester@mps.mpg.de

T. Tadesse
College of Education, Department of Physics Education, Addis Ababa University, Po.Box 1176,
Addis Ababa, Ethiopia

A. Pevtsov
National Solar Observatory, Sunspot, NM 88349, USA
e-mail: apevtsov@nso.edu

1. Introduction

To model and understand the physical mechanisms underlying the various activity phenomena that can be observed in the solar atmosphere, *e.g.*, the onset of flares and coronal mass ejections and the stability of active regions, and to monitor the magnetic helicity and free magnetic energy, the magnetic field vector throughout the atmosphere must be known. However, routine measurements of the solar magnetic field are mainly carried out in the photosphere. The magnetic field in the photosphere is measured using the Zeeman effect of magnetically sensitive solar spectral lines. The problem of measuring the coronal field and its embedded electrical currents thus leads us to use numerical modelling to infer the field strength in the higher layers of the solar atmosphere from the measured photospheric field. Except in eruptions, the magnetic field in the solar corona evolves slowly as it responds to changes in the surface field, implying that the electromagnetic Lorentz forces in this low-β environment are relatively weak and that any electrical currents that exist must be essentially parallel or antiparallel to the magnetic field wherever the field is not negligible.

Due to the low value of the plasma β (the ratio of gas pressure to magnetic pressure), the solar corona is magnetically dominated (Gary, 2001). To describe the equilibrium structure of the static coronal magnetic field when non-magnetic forces are negligible, the force-free assumption is appropriate:

$$(\nabla \times \mathbf{B}) \times \mathbf{B} = 0 \tag{1}$$

$$\nabla \cdot \mathbf{B} = 0 \tag{2}$$

$$\mathbf{B} = \mathbf{B}_{\text{obs}} \quad \text{on photosphere} \tag{3}$$

where $\mathbf{B}$ is the magnetic field and $\mathbf{B}_{\text{obs}}$ is the measured vector field on the photosphere. Equation (1) states that the Lorentz force vanishes (as a consequence of $\mathbf{J} \parallel \mathbf{B}$, where $\mathbf{J}$ is the electric current density) and Equation (2) describes the absence of magnetic monopoles.

The extrapolation methods based on this assumption are termed nonlinear force-free field extrapolation (Sakurai, 1981; Amari *et al.*, 1997; Amari, Boulmezaoud, and Mikic, 1999; Amari, Boulmezaoud, and Aly, 2006; Wu *et al.*, 1990; Cuperman, Demoulin, and Semel, 1991; Demoulin, Cuperman, and Semel, 1992; Inhester and Wiegelmann, 2006; Mikic and McClymont, 1994; Roumeliotis, 1996; Yan and Sakurai, 2000; Valori, Kliem, and Keppens, 2005; Wiegelmann, 2004; Wheatland, 2004; Wheatland and Régnier, 2009; Wheatland and Leka, 2010; Amari and Aly, 2010). For a more complete review of existing methods for computing nonlinear force-free coronal magnetic fields, we refer to the review papers by Amari *et al.* (1997), Schrijver *et al.* (2006), Metcalf *et al.* (2008), and Wiegelmann (2008). Wiegelmann and Neukirch (2006) have developed a code for the self-consistent computation of the coronal magnetic fields and the coronal plasma that uses non-force-free magnetohydrodynamic (MHD) equilibria.

The magnetic field is not force-free in the photosphere, but becomes force-free roughly 400 km above the photosphere (Metcalf *et al.*, 1995). Furthermore, measurement errors, in particular for the transverse field components (*i.e.*, perpendicular to the line of sight of the observer), would destroy the compatibility of a magnetogram with the condition of being force-free. One way to ease these problems is to preprocess the magnetogram data as suggested by Wiegelmann, Inhester, and Sakurai (2006). The preprocessing modifies the boundary values of $\mathbf{B}$ within the error margins of the measurement so that the moduli of force-free integral constraints of Molodensky (1974) are minimised. The resulting boundary values are expected to be more suitable than the original values for an extrapolation into a force-free field.

In the present work, we use a larger computational domain which accommodates most of the connectivity within the coronal region, and we also take the uncertainties of measurements in vector magnetograms into account, as suggested in DeRosa *et al.* (2009). We apply a preprocessing procedure to Synoptic Optical Long-term Investigations of the Sun (SOLIS) data in spherical geometry (Tadesse, Wiegelmann, and Inhester, 2009) by taking account of the curvature of the Sun's surface. For our observations, performed on 28, 29, and 30 March 2008, respectively, the large field of view contains three active regions (ARs: 10987, 10988, 10989).

The full inversion of SOLIS-Vector Spectromagnetograph (VSM) magnetograms yields the magnetic filling factor for each pixel, and it also corrects for magneto-optical effects in the spectral line formation. The full inversion is performed in the framework of the Milne–Eddington (ME) model (Unno, 1956) only for pixels whose polarisation is above a selected threshold. Pixels with polarisation below threshold are left undetermined. These data gaps represent a major difficulty for existing magnetic field extrapolation schemes. Due to the large area of missing data in the example treated here, the reconstructed field model obtained must be treated with some caution. It is very likely that the field strength in the area of missing data was small because the inversion procedure, which calculates the surface field from the Stokes line spectra, abandons the calculation if the signal is below a certain threshold. However, the magnetic field in the corona is dominated by the strongest flux elements on the surface, even if they occupy only a small portion of the surface. Thus we are confident that these dominant flux elements are accounted for in the surface magnetogram, so that the resulting field model is fairly realistic. At any rate, it is the field close to the real, which can be constructed from the available sparse data, so we use a procedure which allows us to incorporate measurement error and to treat regions lacking observational data as in Tadesse *et al.* (2011). The technique has been tested in Cartesian geometry in Wiegelmann and Inhester (2010) for synthetic boundary data.

2. Optimisation Principle in Spherical Geometry

Wheatland, Sturrock, and Roumeliotis (2000) proposed that the variational principle be solved iteratively, which minimises Lorentz forces (Equation (1)) and the divergence of magnetic field (Equation (2)) throughout the volume of interest, V. Later the procedure was improved by Wiegelmann (2004) for Cartesian geometry in a such way that it could use only the bottom boundary on the photosphere as input. Here we use an optimisation approach for the functional ($\mathcal{L}_\omega$) in spherical geometry (Wiegelmann, 2007; Tadesse, Wiegelmann, and Inhester, 2009) and iterate **B** to minimise $\mathcal{L}_\omega$. The modification concerns the input bottom boundary field $\mathbf{B}_{\mathrm{obs}}$, which the model field **B** is not forced to match exactly, but we allow deviations of the order of the observational errors. The modified variational problem is (Wiegelmann and Inhester, 2010; Tadesse *et al.*, 2011):

$$\mathbf{B} = \operatorname{argmin}(\mathcal{L}_\omega)$$

$$\mathcal{L}_\omega = \mathcal{L}_f + \mathcal{L}_d + \nu \mathcal{L}_{\mathrm{photo}} \tag{4}$$

$$\mathcal{L}_f = \int_V \omega_f(r,\theta,\phi) B^{-2} \left| (\nabla \times \mathbf{B}) \times \mathbf{B} \right|^2 r^2 \sin\theta \, dr \, d\theta \, d\phi$$

$$\mathcal{L}_d = \int_V \omega_d(r, \theta, \phi) |\nabla \cdot \mathbf{B}|^2 r^2 \sin\theta \, dr \, d\theta \, d\phi$$

$$\mathcal{L}_{\text{photo}} = \int_S (\mathbf{B} - \mathbf{B}_{\text{obs}}) \cdot \mathbf{W}(\theta, \phi) \cdot (\mathbf{B} - \mathbf{B}_{\text{obs}}) r^2 \sin\theta \, d\theta \, d\phi$$

where $\mathcal{L}_f$ and $\mathcal{L}_d$ measure how well the force-free equation (1) and divergence-free condition (2) are fulfilled, respectively. $\omega_f(r, \theta, \phi)$ and $\omega_d(r, \theta, \phi)$ are weighting functions for the force-free term and divergence-free term, respectively, and are identical for this study. The third integral, $\mathcal{L}_{\text{photo}}$, is a surface integral over the photosphere which relaxes the field on the photosphere towards a force-free solution without too much deviation from the original surface field data, $\mathbf{B}_{\text{obs}}$. In this integral, $\mathbf{W}(\theta, \phi) = \text{diag}(w_{\text{radial}}, w_{\text{trans}}, w_{\text{trans}})$ is a diagonal matrix which gives different weights for observed surface field components depending on its relative accuracy in measurement. In this sense, lack of data is considered most inaccurate and is compensated by setting $W(\theta, \phi)$ to zero in all elements of the matrix.

We use a spherical grid r, θ, ϕ with n_r, n_θ, n_ϕ grid points in the direction of radius, latitude, and longitude, respectively. In the code, we normalise the magnetic field with the average radial magnetic field on the photosphere and the length scale with a solar radius for numerical reasons. Figure 1 shows a map of the radial component of the field as colour-coded with the transverse magnetic field depicted as white arrows. For this particular dataset, about 86% of the data pixels are undetermined. The method works as follows.

- We compute an initial source surface potential field in the computational domain from $\mathbf{B}_{\text{obs}} \cdot \hat{\mathbf{r}}$, the normal component of the surface field at the photosphere at $r = 1 R_\odot$. The

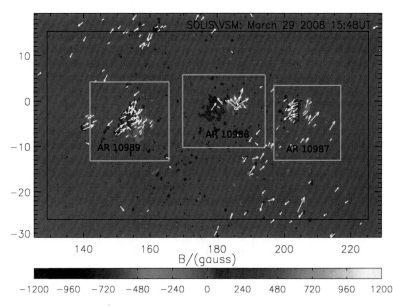

Figure 1 Surface contour plot of radial magnetic field component and vector field plot of transverse field with white arrows. The colour coding shows B_r on the photosphere. The vertical and horizontal axes show latitude, θ (in degrees) and longitude, ϕ (in degrees) on the photosphere. In the area coloured in olive, field values are lacking. The region inside the black box corresponds to the physical domain where the weighting function is unity, and the outside region is the buffer zone where it declines to zero. The blue boxes indicate the domains of the three active regions.

computation is performed by assuming that a currentless ($\mathbf{J} = 0$ or $\nabla \times \mathbf{B} = 0$) approximation holds between the photosphere and some spherical surface S_s (source surface where the magnetic field vector is assumed radial). We compute the solution of this boundary value problem in a standard form of spherical harmonics expansion.

- We minimise $\mathcal{L}_\omega$ (Equations (4)) iteratively. The model magnetic field $\mathbf{B}$ at the surface is gradually driven towards the observed field $\mathbf{B}_{obs}$ while the field in the volume V relaxes to force-free. If the observed field, $\mathbf{B}_{obs}$, is inconsistent, the difference $\mathbf{B} - \mathbf{B}_{obs}$ remains finite depending on the control parameter ν. At data gaps in $\mathbf{B}_{obs}$, the respective field value is automatically ignored.

- The iteration stops when $\mathcal{L}_\omega$ becomes stationary as $\Delta\mathcal{L}_\omega / \mathcal{L}_\omega < 10^{-4}$, where $\Delta\mathcal{L}_\omega$ is the decrease of $\mathcal{L}_\omega$ during an iterative step.

- A convergence to $\mathcal{L}_\omega = 0$ yields a perfect force-free and divergence-free state and exact agreement of the boundary values $\mathbf{B}$ with observations $\mathbf{B}_{obs}$ in regions where the elements of $\mathbf{W}$ are greater than zero. For inconsistent boundary data the force-free and solenoidal conditions can still be fulfilled, but the surface term $\mathcal{L}_{photo}$ will remain finite. This results in some deviation of the bottom boundary data from the observations, especially in regions where w_{radial} and w_{trans} are small. The parameter ν is tuned so that these deviations do not exceed the local estimated measurement error.

3. Results

In this work, we apply our extrapolation scheme to Milne–Eddington inverted vector magnetograph data from the SOLIS survey. As a first step, we remove non-magnetic forces from the observed surface magnetic field using our spherical preprocessing procedure. The code takes $\mathbf{B}_{obs}$ as improved boundary condition.

SOLIS-VSM provides full-disk vector magnetograms, but for some individual pixels the inversion from line profiles to field values may not have been successfully inverted, and field data there will be missing for these pixels (see Figure 1). The different errors for the radial and transverse components of $\mathbf{B}_{obs}$ are accounted for by different values for w_{radial} and w_{trans}. In this work we used $w_{radial} = 20w_{trans}$ for the surface preprocessed fields, as the radial component of $\mathbf{B}_{obs}$ is measured with higher accuracy.

We compute the three-dimensional (3D) magnetic field above the observed surface region inside a wedge-shaped computational box of volume V, which includes an inner physical domain V' and a buffer zone (the region outside the physical domain), as shown in Figure 2. The physical domain V' is a wedge-shaped volume, with two latitudinal boundaries at $\theta_{min} = -26°$ and $\theta_{max} = 16°$, two longitudinal boundaries at $\phi_{min} = 129°$ and $\phi_{max} = 226°$, and two radial boundaries at the photosphere ($r = 1 R_\odot$) and $r = 1.75 R_\odot$. Note that the longitude ϕ is measured from the centre meridian of the back side of the disk. We define V' to be the inner region of V (including the photospheric boundary) with $\omega_f = \omega_d = 1$ everywhere including its six inner boundaries $\delta V'$. We use a position-dependent weighting function to introduce a buffer boundary of $nd = 10$ grid points towards the side and top boundaries of the computational box, V. The weighting functions ω_f and ω_d are chosen to be unity within the inner physical domain V' and decline to 0 with a cosine profile in the buffer boundary region (Wiegelmann, 2004; Tadesse, Wiegelmann, and Inhester, 2009). The framed region in Figure 1 corresponds to the lower boundary of the physical domain V' with a resolution of 114×251 pixels in the photosphere.

The middle panel of Figure 3 shows magnetic field line plots for three consecutive dates of observation. The top and bottom panels of Figure 3 show the position of the three ac-

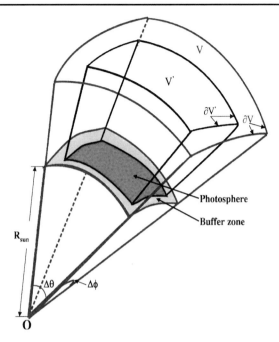

Figure 2 Wedge-shaped computational box of volume V with inner physical domain V' and a buffer zone. O is the centre of the Sun.

tive regions on the solar disk for both the SOLIS full-disk magnetogram[1] and the *Solar and Heliospheric Observatory* (SOHO)/Extreme Ultraviolet Imaging Telescope (EIT)[2] image of the Sun observed at 195 Å on the indicated dates and times. Figure 4 shows some selected magnetic field lines from reconstruction from the SOLIS magnetograms, zoomed in from the middle panels of Figure 3. In each column of Figure 4 the field lines are plotted from the same footpoints to compare the change in topology of the magnetic field over the three-day period of observation. To compare the fields at the three consecutive days quantitatively, we computed the vector correlations between the three field configurations. The vector correlation (C_{vec}) (Schrijver *et al.*, 2006) metric generalises the standard correlation coefficient for scalar functions and is given by

$$C_{vec} = \frac{\sum_i \mathbf{v}_i \cdot \mathbf{u}_i}{\sqrt{\sum_i |\mathbf{v}_i|^2}\sqrt{\sum_i |\mathbf{u}_i|^2}} \tag{5}$$

where $\mathbf{v}_i$ and $\mathbf{u}_i$ are 3D vectors at grid point i. If the vector fields are identical, then $C_{vec} = 1$; if $\mathbf{v}_i \perp \mathbf{u}_i$, then $C_{vec} = 0$. The correlations (C_{vec}) of the 3D magnetic field vectors of 28 and 30 March with respect to the field on 29 March are 0.96 and 0.93, respectively. From these values we can see that there has been no major change in the magnetic field configuration during this period.

We also compute the values of the free magnetic energy estimated from the excess energy of the extrapolated field beyond the potential field satisfying the same $\mathbf{B}_{obs} \cdot \hat{\mathbf{r}}$ boundary condition. Similar estimates have been made by Régnier and Priest (2007) and Thalmann, Wiegelmann, and Raouafi (2008) for single active regions observed at other times. From

---

[1] http://solis.nso.edu/solis_data.html.

[2] http://sohowww.nascom.nasa.gov/data/archive.

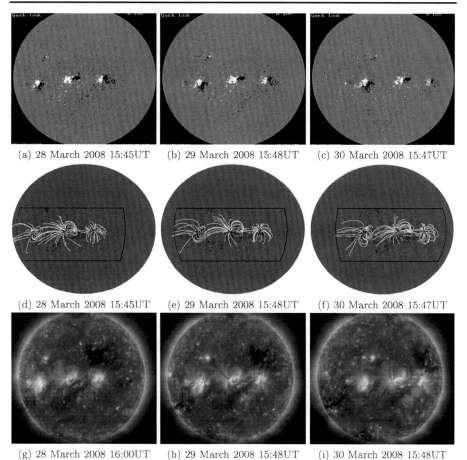

(a) 28 March 2008 15:45UT (b) 29 March 2008 15:48UT (c) 30 March 2008 15:47UT

(d) 28 March 2008 15:45UT (e) 29 March 2008 15:48UT (f) 30 March 2008 15:47UT

(g) 28 March 2008 16:00UT (h) 29 March 2008 15:48UT (i) 30 March 2008 15:48UT

Figure 3 Top row: SOLIS-VSM magnetograms of respective dates. Middle row: Magnetic field lines reconstructed from magnetograms on the top panel. Bottom row: EIT image of the Sun at 195 Å on indicated dates.

Table 1 The magnetic energy associated with extrapolated NLFF field configurations for the three particular dates.

Date	$E_{\text{nlff}}(10^{32}$ erg)	$E_{\text{pot}}(10^{32}$ erg)	$E_{\text{free}}(10^{32}$ erg)
28 March 2008	57.34	53.89	3.45
29 March 2008	57.48	54.07	3.41
30 March 2008	57.37	53.93	3.44

the corresponding potential and force-free magnetic field, $\mathbf{B}_{\text{pot}}$ and $\mathbf{B}$, respectively, we can estimate an upper limit to the free magnetic energy associated with coronal currents

$$E_{\text{free}} = E_{\text{nlff}} - E_{\text{pot}} = \frac{1}{8\pi} \int_{V'} \left(B_{\text{nlff}}^2 - B_{\text{pot}}^2 \right) r^2 \sin\theta \, dr \, d\theta \, d\phi. \tag{6}$$

The computed energy values are listed in Table 1. The free energy on all three days is about 3.5×10^{32} erg. The magnetic energy associated with the potential field configuration is

(a) 28 March 2008 15:45UT

(b) 29 March 2008 15:48UT

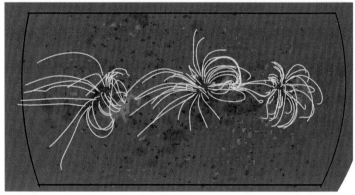

(c) 30 March 2008 15:47UT

Figure 4 Some magnetic field line plots reconstructed from SOLIS magnetograms using nonlinear force-free modelling. The colour coding shows B_r on the photosphere.

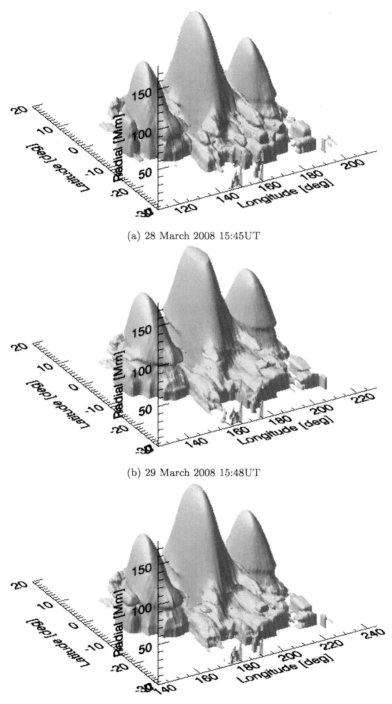

(a) 28 March 2008 15:45UT

(b) 29 March 2008 15:48UT

(c) 30 March 2008 15:47UT

Figure 5 Iso-surfaces (ISs) of the absolute nonlinear force-free (NLFF) magnetic energy density $(7.5 \times 10^{16}$ erg) for the three consecutive dates computed within the entire computational domain.

Table 2 The percentage of the total magnetic flux shared between the three ARs. Φ_{11}, Φ_{22}, and Φ_{33} denote magnetic flux of AR 10989 (left), AR 10988 (middle), and AR 10987 (right) of Figure 1, respectively.

$\Phi_{\alpha\beta}$	28th			29th			30th		
	$\alpha = 1$	2	3	$\alpha = 1$	2	3	$\alpha = 1$	2	3
$\beta = 1$	56.37	5.59	0.00	56.50	5.48	0.00	56.50	5.48	0.00
2	13.66	81.12	1.43	13.66	81.22	1.43	13.66	81.22	2.22
3	0.00	0.48	71.47	0.00	0.48	71.80	0.00	0.48	71.80
Elsewhere	29.97	12.82	27.10	29.84	12.82	26.77	29.84	12.82	25.98

Table 3 The percentage of the total electric current shared between the three ARs. I_{11}, I_{22}, and I_{33} denote electric current of AR 10989 (left), AR 10988 (middle), and AR 10987 (right) of Figure 1, respectively.

$I_{\alpha\beta}$	28th			29th			30th		
	$\alpha = 1$	2	3	$\alpha = 1$	2	3	$\alpha = 1$	2	3
$\beta = 1$	82.47	0.19	0.00	86.36	0.19	0.00	94.16	0.19	0.00
2	0.65	85.25	1.42	0.65	85.25	1.42	0.65	85.25	3.55
3	0.00	0.38	82.27	0.00	0.38	82.27	0.00	0.38	82.27
Elsewhere	16.88	14.18	16.31	12.99	14.18	16.31	5.19	14.18	14.18

about 54×10^{32} erg. Hence E_{nlff} exceeds E_{pot} by only 6%. Figure 5 shows iso-surface plots of magnetic energy density in the volume above the active regions. There are strong energy concentrations above each active region. There were no major changes in the magnetic energy density over the observation period, and there was no major eruptive phenomenon during those three days in the region observed.

In our previous work (Tadesse *et al.*, 2011), we have studied the connectivity between two neighbouring active regions. In this work with an even larger field of view, the three active regions share a decent amount of magnetic flux compared to their internal flux from one polarity to the other (see Figure 4). In terms of the electric current they are much more isolated. In order to quantify these connectivities, we have calculated the magnetic flux and the electric currents shared between active regions. For the magnetic flux, *e.g.*, we use

$$\Phi_{\alpha\beta} = \sum_i |\mathbf{B}_i \cdot \hat{r}| R_\odot^2 \sin(\theta_i) \Delta\theta_i \Delta\phi_i \tag{7}$$

where the summation is over all pixels of AR_α from which the field line ends in AR_β or $i \in \text{AR}_\alpha \| \text{conjugate footpoint}(i) \in \text{AR}_\beta$. The indices α and β denote the active regions and the index number 1 corresponds to AR 10989, 2 to AR 10988, and 3 to AR 10987 of Figure 1. For the electric current we replace the magnetic field, $\mathbf{B}_i \cdot \hat{r}$, by the vertical current density $\mathbf{J}_i \cdot \hat{r}$ in Equation (7). Whenever the end point of a field line falls outside (blue rectangles in Figure 1) the three ARs, we categorise it as ending elsewhere. Tables 2 and 3, respectively, show the percentage of the total magnetic flux and electric current shared between the three ARs. For example, the first column of Table 2 shows that 56.37% of positive polarity of AR_1 is connected to negative polarity of AR_1; line 2 shows that 13.66% of positive/negative polarity of AR_1 is connected to positive/negative polarity of AR_2, and line 3 shows that there are no field lines (0%) connecting positive/negative polarity of AR_1 with positive/negative polarity of AR_3. The same technique applies for Table 3. The three active regions are magnetically connected but much less so by electric currents.

4. Conclusions

We have investigated the coronal magnetic field associated with three ARs: 10987, 10987, and 10989, on 28, 29, and 30 March 2008 by analysing SOLIS-VSM data. We used an optimisation method for the reconstruction of NLFF coronal magnetic fields in spherical geometry by restricting the code to limited parts of the Sun (Wiegelmann, 2007; Tadesse, Wiegelmann, and Inhester, 2009; Tadesse *et al.*, 2011). The code was modified so that it allows us to deal with lacking data and regions with poor signal-to-noise ratio in a systematic manner (Wiegelmann and Inhester, 2010; Tadesse *et al.*, 2011).

We have studied the time evolution of the magnetic field over the three-day period and found no major changes in topologies, as there was no major eruption event. The magnetic energies calculated in the large wedge-shaped computational box above the three ARs were not far apart in value. This is the first study which contains three well-separated ARs in our model. It was made possible by the use of spherical coordinates, allowing us to analyse linkage between the ARs. The ARs share a decent amount of magnetic flux compared to their internal flux from one polarity to the other. However, in terms of electric current they are much more isolated.

Acknowledgements   SOLIS-VSM vector magnetograms are produced cooperatively by the National Science Foundation (NSF)/National Solar Observatory (NSO) and the National Aeronautics and Space Administration (NASA)/Living With a Star (LWS). The NSO is operated by the Association of Universities for Research in Astronomy, Inc., under cooperative agreement with the NSF. Tilaye Tadesse Asfaw acknowledges a fellowship of the International Max-Planck Research School at the Max-Planck Institute for Solar System Research. The work of T. Wiegelmann was supported by DLR-grant 50 OC 0501.

References

Amari, T., Aly, J.: 2010, *Astron. Astrophys.* **522**, A52.
Amari, T., Boulmezaoud, T.Z., Mikic, Z.: 1999, *Astron. Astrophys.* **350**, 1051.
Amari, T., Boulmezaoud, T.Z., Aly, J.J.: 2006, *Astron. Astrophys.* **446**, 691.
Amari, T., Aly, J.J., Luciani, J.F., Boulmezaoud, T.Z., Mikic, Z.: 1997, *Solar Phys.* **174**, 129.
Cuperman, S., Demoulin, P., Semel, M.: 1991, *Astron. Astrophys.* **245**, 285.
Demoulin, P., Cuperman, S., Semel, M.: 1992, *Astron. Astrophys.* **263**, 351.
DeRosa, M.L., Schrijver, C.J., Barnes, G., Leka, K.D., Lites, B.W., Aschwanden, M.J., *et al.*: 2009, *Astrophys. J.* **696**, 1780.
Gary, G.A.: 2001, *Solar Phys.* **203**, 71.
Inhester, B., Wiegelmann, T.: 2006, *Solar Phys.* **235**, 201.
Metcalf, T.R., Jiao, L., McClymont, A.N., Canfield, R.C., Uitenbroek, H.: 1995, *Astrophys. J.* **439**, 474.
Metcalf, T.R., Derosa, M.L., Schrijver, C.J., Barnes, G., van Ballegooijen, A.A., Wiegelmann, T., Wheatland, M.S., Valori, G., McTiernan, J.M.: 2008, *Solar Phys.* **247**, 269.
Mikic, Z., McClymont, A.N.: 1994, In: Balasubramaniam, K.S., Simon, G.W. (eds.) *Solar Active Region Evolution: Comparing Models with Observations*, ASP Conf. Ser. **68**, 225.
Molodensky, M.M.: 1974, *Solar Phys.* **39**, 393.
Régnier, S., Priest, E.R.: 2007, *Astrophys. J. Lett.* **669**, 53.
Roumeliotis, G.: 1996, *Astrophys. J.* **473**, 1095.
Sakurai, T.: 1981, *Solar Phys.* **69**, 343.
Schrijver, C.J., Derosa, M.L., Metcalf, T.R., Liu, Y., McTiernan, J., Régnier, S., Valori, G., Wheatland, M.S., Wiegelmann, T.: 2006, *Solar Phys.* **235**, 161.
Tadesse, T., Wiegelmann, T., Inhester, B.: 2009, *Astron. Astrophys.* **508**, 421.
Tadesse, T., Wiegelmann, T., Inhester, B., Pevtsov, A.: 2011, *Astron. Astrophys.* **527**, A30.
Thalmann, J.K., Wiegelmann, T., Raouafi, N.E.: 2008, *Astron. Astrophys.* **488**, L71.
Unno, W.: 1956, *Publ. Astron. Soc. Japan* **8**, 108.
Valori, G., Kliem, B., Keppens, R.: 2005, *Astron. Astrophys.* **433**, 335.
Wheatland, M.S.: 2004, *Solar Phys.* **222**, 247.
Wheatland, M.S., Leka, K.D.: 2010, ArXiv e-prints.

Wheatland, M.S., Régnier, S.: 2009, *Astrophys. J. Lett.* **700**, 88.

Wheatland, M.S., Sturrock, P.A., Roumeliotis, G.: 2000, *Astrophys. J.* **540**, 1150.

Wiegelmann, T.: 2004, *Solar Phys.* **219**, 87.

Wiegelmann, T.: 2007, *Solar Phys.* **240**, 227.

Wiegelmann, T.: 2008, *J. Geophys. Res.* **113**, 3.

Wiegelmann, T., Inhester, B.: 2010, *Astron. Astrophys.* **516**, A107.

Wiegelmann, T., Neukirch, T.: 2006, *Astron. Astrophys.* **457**, 1053.

Wiegelmann, T., Inhester, B., Sakurai, T.: 2006, *Solar Phys.* **233**, 215.

Wu, S.T., Sun, M.T., Chang, H.M., Hagyard, M.J., Gary, G.A.: 1990, *Astrophys. J.* **362**, 698.

Yan, Y., Sakurai, T.: 2000, *Solar Phys.* **195**, 89.

Solar Phys (2012) 277:131–151
DOI 10.1007/s11207-011-9830-6

Magnetic Energy Storage and Current Density Distributions for Different Force-Free Models

S. Régnier

Received: 4 January 2011 / Accepted: 20 July 2011 / Published online: 4 October 2011
© Springer Science+Business Media B.V. 2011

Abstract In the last decades, force-free-field modelling has been used extensively to describe the coronal magnetic field and to better understand the physics of solar eruptions at different scales. Especially the evolution of active regions has been studied by successive equilibria in which each computed magnetic configuration is subject to an evolving photospheric distribution of magnetic field and/or electric-current density. This technique of successive equilibria has been successful in describing the rate of change of the energetics for observed active regions. Nevertheless the change in magnetic configuration due to the increase/decrease of electric current for different force-free models (potential, linear and nonlinear force-free fields) has never been studied in detail before. Here we focus especially on the evolution of the free magnetic energy, the location of the excess of energy, and the distribution of electric currents in the corona. For this purpose, we use an idealised active region characterised by four main polarities and a satellite polarity, allowing us to specify a complex topology and sheared arcades to the coronal magnetic field but no twisted flux bundles. We investigate the changes in the geometry and connectivity of field lines, the magnetic energy and current-density content as well as the evolution of null points. Increasing the photospheric current density in the magnetic configuration does not dramatically change the energy-storage processes within the active region even if the magnetic topology is slightly modified. We conclude that for reasonable values of the photospheric current density (the force-free parameter $\alpha < 0.25$ Mm^{-1}), the magnetic configurations studied do change but not dramatically: *i*) the original null point stays nearly at the same location, *ii*) the field-line geometry and connectivity are slightly modified, *iii*) even if the free magnetic energy is significantly increased, the energy storage happens at the same location. This extensive study of different force-free models for a simple magnetic configuration shows that some topological elements of an observed active region, such as null points, can be reproduced with confidence only by considering the potential-field approximation. This study is a prelimi-

Solar Flare Magnetic Fields and Plasmas
Guest Editors: Y. Fan and G.H. Fisher

S. Régnier (✉)
Jeremiah Horrocks Institute, University of Central Lancashire, Preston, Lancashire PR1 2HE, UK
e-mail: SRegnier@uclan.ac.uk

nary work aiming at understanding the effects of electric currents generated by characteristic photospheric motions on the structure and evolution of the coronal magnetic field.

Keywords Magnetic fields, corona · Active regions, structure · Electric currents and current sheets

1. Introduction

One key unsolved issue in solar physics is the generation and effects of electric currents from photospheric motions on the stressed and sheared coronal magnetic field. With the development of reliable techniques such as coronal magnetic-field extrapolations based on complex distributions of photospheric currents (*e.g.*, reviews by Régnier, 2007; Wiegelmann, 2008), it is important to understand how a modelled magnetic-field configuration is subject to change due to slight modifications of the photospheric-current distribution and thus the difference between several force-free assumptions using the same boundary conditions. The main aim is to understand these changes in the geometry and topology of field lines assuming that the magnetic field is in a force-free equilibrium. In Régnier and Priest (2007), we have performed a first comparison between four different active regions with different behaviours due to the complex distribution of the photospheric field and of the electric currents representing the history of the evolution of the active region. This comparison was done for the same force-free model, namely the nonlinear force-free field. We showed that, statistically speaking, the magnetic-field lines are longer and higher in a nonlinear force-free field compared to the corresponding potential field. To develop our understanding of the effects of electric currents on magnetic configurations, Régnier (2009) compared the behaviour of a simple bipolar field subject to different distributions of electric current using different force-free models. The bipolar field has been studied in terms of magnetic-energy storage and magnetic-helicity changes. We showed that the amount of electric currents that can be injected in a magnetic-equilibrium configuration depends strongly on the spatial distribution of the currents, the existence of return currents having a stabilising effect on the magnetic configuration. The next step developed in this article is to understand the effects of electric currents on a magnetic configuration having predominant topological elements (*i.e.* a null point in the domain of interest).

In the past decades, magnetic topology has become a key ingredient in understanding the origin of flares in active regions.

The general definition of magnetic topology concerns the properties of magnetic-field lines and magnetic-flux surfaces that are invariant under continuous deformation in plasma conditions satisfying the frozen-in assumption (Low, 2006, 2007; Janse, Low, and Parker, 2010; Berger and Prior, 2006). This definition implies that the number of null points does not change and the connectivity of field lines rigidly anchored to the boundaries of the domain is also invariant under the above conditions. However, we focus here on the evolution of topological elements and connectivity of field lines obtained from force-free models: the field lines are not rigidly anchored (the models allow for reconnection of field lines) in order to maintain a force-free state. Since the magnetic topology is not required to be preserved, we restrict the definition of the magnetic topology for a given equilibrium state to the ensemble of topological elements forming the magnetic skeleton (Bungey, Titov, and Priest, 1996). Despite an extensive literature on magnetic topology, the theoretical background in three dimensions was developed only recently (see review by Priest and Forbes, 2000). The 3D topological elements constituting the skeleton of a magnetic configuration can be divided into two parts: the true topology containing null points, separatrix surfaces, and separators, and the quasi-topology including quasi-separatrix layers and hyperbolic flux tubes (and the true topology). We focus especially on the location and properties of null points as a proxy for describing the topology of magnetic configuration. This study aims at understanding the possible changes of properties and location of null points subject to the continuous variation of a free parameter in various force-free models.

It has been proven that the topological elements of a coronal magnetic configuration are of prime importance to study the onset of flares and coronal mass ejections (CMEs). For instance, in the classical model of a flare, magnetic reconnection occurs at a null point or in a current sheet formed along a topological element. Combining observations and coronal-field models, Aulanier *et al.* (2000) have shown that a powerful flare associated with a CME involves a coronal null point and a spine field line. This topological study in conjunction with EUV observations supports the breakout model (Antiochos, Devore, and Klimchuk, 1999) as a triggering mechanism for this particular event. Recently, Zhao *et al.* (2008) have derived the skeleton of an active region and its temporal evolution before and after an eruptive event. The authors have found several coronal null points in a quasi-force-free field configuration. Unfortunately, the null points found by Zhao *et al.* (2008) did not satisfy the properties of null points for force-free fields and for divergence-free fields (see Appendix A). Other topological studies have been carried out to better understand the release of magnetic energy and the reconnection processes in active region evolution (Démoulin, Hénoux, and Mandrini, 1994; Deng *et al.*, 2005; Barnes, Longcope, and Leka, 2005; Régnier and Canfield, 2006; Li *et al.*, 2006; Luoni *et al.*, 2007; Barnes, 2007), during blinkers (Subramanian *et al.*, 2008), and in the quiet Sun (Schrijver and Title, 2002; Close, Heyvaerts, and Priest, 2004; Régnier, Parnell, and Haynes, 2008)

Our work has been motivated by two earlier articles on this topic by Démoulin, Hénoux, and Mandrini (1994) and Hudson and Wheatland (1999). Démoulin, Hénoux, and Mandrini (1994) have compared the topology of potential and linear force-free fields for a quadrupolar configuration with a coronal null. They found that the topology is similar for the point-charge model but they also noticed that for a bipolar model (based on extended sources) another null can be created in the linear force-free configuration for large value of the force-free parameter [α]. Based on the point-charge model, Bungey, Titov, and Priest (1996) reached the same conclusion for different charge distributions. Hudson and Wheatland (1999) have considered a symmetric quadrupolar distribution to study the connectivity of field lines for potential, linear, and nonlinear force-free fields. They concluded that the topology can be drastically different from one model to the other. For non-symmetric cases, Brown and Priest (2000) have found that the topology of a force-free configuration can be similar using the point-charge model. As a step forward to the understanding of the magnetic topology of reconstructed coronal fields, we carry out a comparison between different models of magnetic fields (potential, linear and nonlinear force-free fields) for a configuration having a coronal null point assuming a continuous distribution of magnetic field at the bottom boundary and no symmetry. We describe the changes in the magnetic configurations based on the evolution of the geometry and connectivity of magnetic-field lines and in terms of distribution of magnetic energy and electric currents. In addition, we study the properties of null points as a proxy of the complexity of the field, keeping in mind that the separators and separatrices play an important role in the release of magnetic energy (*e.g.*, Priest, Longcope, and Heyvaerts, 2005). It is worth noticing that we only focus on the modelling of magnetic-field equilibria and their changes through the increase of electric currents, whilst recently Santos, Büchner, and Otto (2011) have described the changes in magnetic fields subject to characteristic photospheric motions using a magnetohydrodynamic (MHD) approach. They have found that the topology of a quadrupolar magnetic field remains stable

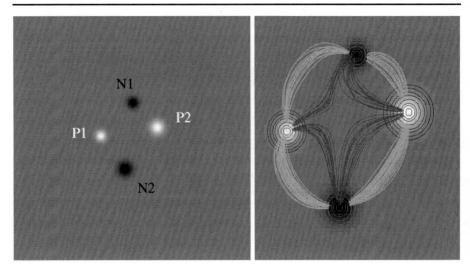

Figure 1 (Left) Quadrupolar distribution of the vertical component of the magnetic field [B_z] used as initial field (black and white are negative (N1 and N2) and positive (P1 and P2) polarities). The total magnetic flux is balanced. (Right) Close-up of a few field lines (red and green) depicting the geometry and topology of the potential-field configuration. White polarities and red contours (black polarities and blue contours) are positive (negative) values of B_z.

whatever the perturbations imposed even in the case of strong currents. The authors claimed that their results can easily be generalised to more complex magnetic-field distributions and non-generic and symmetric cases. Despite their sophisticated MHD approach, they do not study the changes in magnetic energy or in the connectivity of field lines.

In Section 2.1, we construct a magnetic configuration with a coronal null point from which we will reconstruct the different models (see Section 2.2). We thus analyse the geometry of field lines in Section 3 and their connectivity in Section 4. The change in topology in the different models is discussed in Section 5. In addition, we study how the magnetic energy is stored (Section 6) and the electric currents are concentrated (Section 7). In Section 8, we discuss the implications for future topological studies from reconstructed magnetic fields.

2. Constructing a Coronal Null Point

2.1. The Potential-Field Distribution

We first build a quadrupolar distribution of the vertical component of the magnetic field [B_z] at the bottom boundary (see Figure 1 left). The polarities are defined as Gaussian distributions with the field strength at the centre and the width of the distribution as free parameters: N1 and N2 (P1 and P2) are negative (positive) polarities. To realistically model a solar active region, we assume that the spatial resolution is 1 Mm giving a characteristic size of 140 Mm. The four polarities are placed such that there is no symmetry. Each polarity has a maximum field strength of 2000 G in absolute value and a different width of the Gaussian distribution. The total magnetic flux is balanced.

From the magnetogram depicted in Figure 1 left, we compute the potential field in the coronal volume [Ω: 140 pixels × 140 pixels × 120 pixels] imposing closed boundary conditions on the sides and top of the computational box. A few magnetic field lines have been

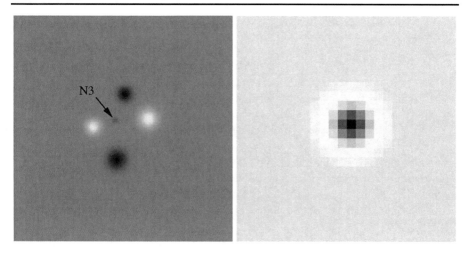

Figure 2 (Left) Distribution of the vertical component of the magnetic field at the bottom boundary (black and white for negative and positive polarities); (Right) Typical ring distribution of J_z that we impose on the positive polarities.

selected in Figure 1 right to show the geometry and topology of the potential field associated with this quadrupolar distribution. There is no null point in the potential-field configuration. However, the magnetic configuration has a topology characterised by quasi-separatrix layers (QSLs) as shown in Figure 1 right, by the red field lines dividing the domain in four distinct regions. The study of the QSLs is beyond the scope of this article. We note that the simplest magnetic configuration with a topology is a configuration with three polarities as studied in detail by Brown and Priest (1999). We have chosen a quadrupolar configuration in order to confine the null point inside the strong-field region.

To create a coronal null point, we emerge a polarity N3 (negative in this experiment) at the location where there is a field-strength minimum of the quadrupolar distribution (see Figure 2 left). The total magnetic flux is kept balanced. We then compute the potential field associated with this new magnetogram, and with the same boundary conditions as in the quadrupolar case. The height of the null point depends on the field strength of the polarity N3: for a maximum field strength of –800 G, the null point is located 6.9 Mm above the bottom boundary (see also Table 1). In the following experiment, we will use the vertical component of the magnetic field depicted in Figure 2 left as a boundary condition for the different force-free models.

2.2. The Magnetic-Field Models

2.2.1. The Grad–Rubin Algorithm

The computed linear and nonlinear force-free fields are based on the Grad and Rubin (1958) numerical scheme described by Amari *et al.* (1997, 1999) and Amari, Boulmezaoud, and Aly (2006). The same boundary conditions are used for all the models: the vertical magnetic-field component everywhere and the distribution of α in one chosen polarity on the bottom boundary, and closed boundary conditions on the sides and top of the computational box. Using the same boundary conditions for all models allows us to perform a direct comparison of the magnetic energy and the topology of the different magnetic configurations.

The force-free field in the volume above the bottom boundary is thus governed by the following equations:

$$\nabla \times \mathbf{B} = \alpha \mathbf{B}, \tag{1}$$

$$\mathbf{B} \cdot \nabla \alpha = 0, \tag{2}$$

$$\nabla \cdot \mathbf{B} = 0, \tag{3}$$

where $\mathbf{B}$ is the magnetic-field vector in the domain Ω above the photosphere [$\delta\Omega$], and α is a function of space defined as the ratio of the vertical current density [J_z], and the vertical magnetic field component [B_z]. From Equation (2), α is constant along a field line. In terms of the magnetic field $\mathbf{B}$, the Grad–Rubin iterative scheme can be written as follows:

$$\mathbf{B}^{(n)} \cdot \nabla \alpha^{(n)} = 0 \quad \text{in } \Omega, \tag{4}$$

$$\alpha^{(n)}|_{\delta\Omega^{\pm}} = h, \tag{5}$$

where $\delta\Omega^{\pm}$ is defined as the domain on the photosphere for which B_z is positive (+) or negative (−) and,

$$\nabla \times \mathbf{B}^{(n+1)} = \alpha^{(n)} \mathbf{B}^{(n)} \quad \text{in } \Omega, \tag{6}$$

$$\nabla \cdot \mathbf{B}^{(n+1)} = 0 \quad \text{in } \Omega, \tag{7}$$

$$B_z^{(n+1)}|_{\delta\Omega} = g, \tag{8}$$

$$\lim_{|r| \to \infty} |\mathbf{B}| = 0. \tag{9}$$

The boundary conditions on the photosphere are given by the distribution [g] of B_z on $\delta\Omega$ (see Equation (8)) and by the distribution [h] of α on $\delta\Omega$ for a given polarity (see Equation (5)). We also impose that

$$B_n = 0 \quad \text{on } \Sigma - \delta\Omega \tag{10}$$

where Σ is the surface of the computational box, n refers to the component normal to the surface. These conditions mean that no field line can enter or leave the computational box. To ensure the latter condition, we have chosen a bottom boundary large enough for the magnetic-field strength to tend to zero near the edges of the field-of-view.

2.2.2. Linear Force-Free Fields

The linear force-free models are based on the Grad–Rubin algorithm where the distribution of α is a constant. We choose values of α ranging from −1 to 1 Mm^{-1} with a step $\delta\alpha = 0.02$ Mm^{-1}. These α values correspond to active-region values reported, for instance, by Leka and Skumanich (1999) and computed by assuming that the measured photospheric field is force-free.

2.2.3. Nonlinear Force-Free Fields

In addition to the vertical component of the magnetic field, the Grad–Rubin scheme requires a distribution for the current density [or α] in order to derive the nonlinear force-free field. Our choice goes to the so-called *ring* distribution defined by a second-order Hermite polynomial function as follows:

$$J_z = 2J_{z0}[r^2 - C_0]\exp\left(-\frac{r^2}{\sigma^2}\right), \tag{11}$$

where r is measured from the centre of the source and C_0 is a constant that ensures a zero net current. An example of a ring distribution is depicted in Figure 2 right. For the sake of comparison, we choose J_{z0} [$-20, -10, 10, 20$] mA m^{-2}, which are characteristic values of the current density in active regions as has been measured in the photosphere from vector magnetograms (*e.g.*, Leka and Skumanich, 1999). The distribution of α is then given by

$$\alpha = \frac{\mu_0 J_z}{B_z}. \tag{12}$$

In accordance with the Grad–Rubin mathematically well-posed boundary-value problem, we impose the distribution of α in one chosen polarity (the positive polarities in this experiment) as boundary condition for the nonlinear force-free field. We use the same grid and the same side and top boundary conditions as for the potential and linear force-free fields.

This particular choice of the vertical current distribution is justified by the study reported by Régnier (2009), which analysed the behaviour of a simple bipolar field under the assumption of a nonlinear force-free field by using several distribution of J_z (or α). This study showed that the ring distribution of current gives magnetic configurations that are more stable, and in which a large amount of current can be injected. The conclusion is easily explained by the stabilising effects of the return currents. Ring-current distributions have been used before for simulating twisted flux tubes in MHD models (Magara and Longcope, 2003).

3. Statistical Study of Field-Line Geometry

To study the differences between the different magnetic configurations obtained for each model, the first step is to analyse the changes in the geometry of field lines. So we first select field lines by considering their footpoints with a field strength above 100 G in absolute value. In the following, we will focus on four different distributions, which are characteristic of this study: the potential field, two linear force-free field configurations for $\alpha = -0.62$ and 0.74 Mm^{-1}, and one nonlinear force-free field with $J_{z0} = 20$ mA m^{-2}.

In Figure 3, the scatter plots of the length and height of the selected field lines are shown for the four configurations. It is clear that the injection of electric currents in a magnetic configuration changes, in a statistical sense, the geometry of field lines compared to the potential field but not the same way for each model. The changes in geometry are more pronounced for the linear force-free models whilst the geometry remains similar for the nonlinear force-free field. As mentioned above, the similarity of the nonlinear force-free model and the potential field and the differences with the linear force-free fields are due to the existence of return currents and their stabilising effects. We now study more closely how the changes in the geometry of field lines evolve when the current density is increased. In

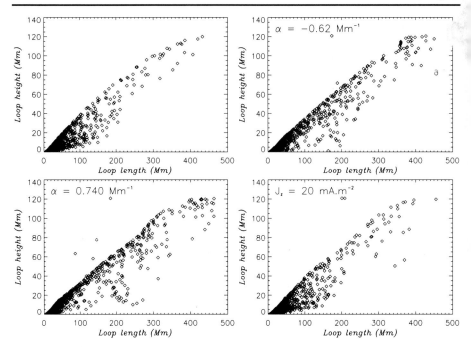

Figure 3 Scatter plots of the length and height of selected field lines for (a) the potential field, (b) and (c) the linear force-free field with $\alpha = -0.62$ and 0.74 Mm^{-1} respectively, and (d) the nonlinear force-free field with $J_{z0} = 20$ mA m^{-2}.

Figure 4, we plot the cumulative-distribution functions of the length (left) and height (right) of selected field lines. We determine that 50% of the field lines are shorter than 25 Mm and lower than 5 Mm for the potential field, whilst 50% of the field lines reaches 50 Mm in length and 15 Mm in height for $\alpha = 1$ Mm^{-1}. The differences between the different linear force-free models occur for field lines longer than 50 Mm and higher than 10 Mm. We also notice that the main differences occur for values of α greater than 0.25 Mm^{-1}. The CDFs are consistent with the analysis of Régnier and Priest (2007): by studying the force-free-field configurations of four different active regions, the authors have concluded that, statistically, the field lines in a nonlinear force-free field are longer and higher than for the corresponding potential-field configuration. We will later refer to values of α less than 0.25 Mm^{-1} as reasonable values of α.

4. Field-Line Connectivity

To depict the connectivity of the field lines, we associate, on a 2D map, the location of the footpoint of a field line to its length. The connectivity plots are drawn for the four characteristic force-free fields in Figure 5. We first notice that the connectivity of the parasitic polarity is not modified for the different models. The changes of connectivity mostly affect the field lines on the outer edges of the polarities and not the regions of strong magnetic-field concentrations. It is worth noticing that the Grad–Rubin algorithm computes the nonlinear force-free field from the positive polarities. For the linear force-free fields, we notice that the long field lines are moved counter-clockwise (clockwise) when the negative (positive)

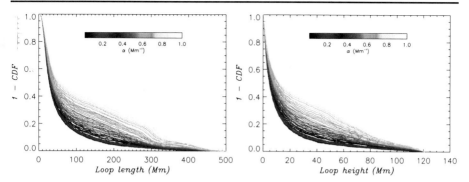

Figure 4 Cumulative distribution function [CDF] for the linear force-free models (coloured curves for the different values of α) for the field-line length (left) and the field line height (right). We only plot the distribution for positive values of α to avoid confusion.

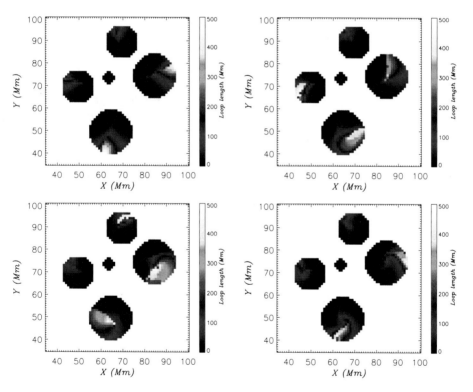

Figure 5 Connectivity maps for a restricted field of view ($x = [30, 110]$, $y = [30, 110]$) depicting the field lines having the same length for the potential field (top left), the force-free field with $\alpha = -0.62$ Mm^{-1} (top right) and $\alpha = 0.74$ Mm^{-1} (bottom left), and the nonlinear force-free field with $J_{z0} = 20$ mA m^{-2} (bottom right). The colour bar indicates the loop lengths in Mm.

values of α are increased in absolute value. For the nonlinear force-free field, a positive (negative) value of J_{z0} gives the same behaviour as a negative (positive) value of α in the linear force-free configurations. Therefore the connectivity of field lines depends strongly on the amount of electric current injected in the magnetic configurations.

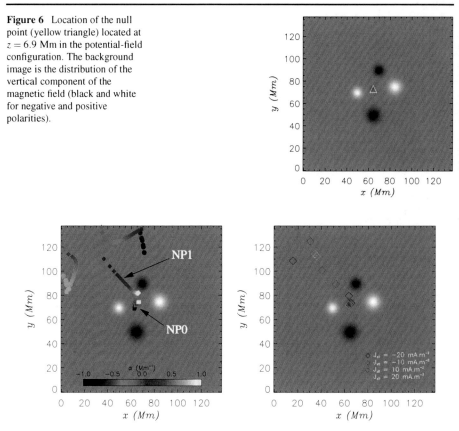

Figure 6 Location of the null point (yellow triangle) located at $z = 6.9$ Mm in the potential-field configuration. The background image is the distribution of the vertical component of the magnetic field (black and white for negative and positive polarities).

Figure 7 Same as Figure 6: (Left) linear force-free fields with α varying from -1 to 1 Mm^{-1}. NP0 is depicted by a square symbol whilst other null points are depicted by a diamond (circle) symbol when their height is greater (lower) than NP0. (Right) nonlinear force-free fields with $J_{z0} = [-20, -10, 10, 20]$ mA m^{-2} (diamonds).

5. Null Point Properties

5.1. Location of Null Points

We locate the null points within the magnetic configurations using the trilinear interpolation method developed by Haynes and Parnell (2007). We then compare the location of the null points for the different models. We plot the location of the null points onto the x–y-plane for the potential field in Figure 6, for the linear force-free fields with α between -1 and 1 Mm^{-1} in Figure 7 left, and for the nonlinear force-free fields using several values of J_{z0} in Figure 7 right. There is only one null point NP0 in the potential field located 6.9 Mm above the parasitic negative polarity (see Figure 6). In Figure 7, the null points present in the magnetic configuration for both the linear and nonlinear force-free fields can be divided into two groups:

i) Near the location of the potential null point, a null point is found for all models whatever the value of the current density or the force-free parameter α.

ii) Other null points can appear mostly near the boundaries but also in strong field regions.

Figure 8 Height of the null points in the potential and linear force-free field models for α varying from -1 to 1 Mm^{-1}. The dashed line indicates the height of NP0 in the potential configuration. Triangles (diamonds) indicate positive (negative) null points.

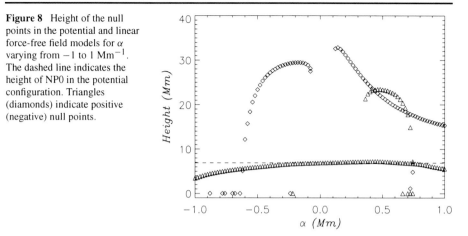

For the first group, we conclude that the null point NP0 created in the potential field is stable in the other models, and its location is just slightly influenced by the current density: the null point is moving up and down, left and right depending on the sign of the current density. For the second group, we can already notice that most of the null points are located near the side boundaries, and in addition the null point NP1 is moving towards strong-field regions when α is increased (see Figure 7 left). However, we need to investigate the properties of the null point to draw conclusions (see Section 5.2).

In Figure 8, we plot the distribution of null points as a function of height for the linear force-free configurations. We indicate the sign of the null points: triangles (diamonds) for positive (negative) null points. This plot allows us to track the null points depending on the value of the parameter α and on the sign of the null points. We notice that the potential-field null point NP0 evolves smoothly when the α parameter (in absolute value) increases: the null point height varies continuously from 4 Mm to 7.5 Mm whilst the null point height in the potential field is 6.9 Mm as indicated by the dashed line in Figure 8 (see also Table 1).

Other null points appear when $|\alpha| > 0.1$ Mm^{-1} (see Figure 8). We obtain up to five null points at $\alpha \approx 0.7$ Mm^{-1} (see Table 1). Several null points are located near the bottom boundary. All null points are at a height less than 40 Mm (one-third of the vertical length of the computational box) where the bipolar field is dominant: the complexity of the quadrupolar and parasitic polarities is located below 40 Mm. Note that the bipolarisation of the magnetic field and its associated height are also measures of the complexity of the magnetic field.

The height of the stable null point decreases when α is negative and it increases when α is positive. Nevertheless for $\alpha > 0.5$ Mm^{-1}, the height starts to decrease, influenced by the null point NP1 propagating in the strong field region towards NP0 (see Figure 7 left).

In Figure 7 right, we find the same groups of null points for the nonlinear force-free fields. We note that, for this experiment, a nonlinear force-free field with negative (positive) values of J_{z0} has the same behaviour as a linear force-free field with positive (negative) values of α.

5.2. Properties

As discussed in Appendix A, the null points can be classified as negative and positive depending on whether the fan field lines are radiating in or out, respectively, from the null point. To derive the spectral properties of a null point, we first need to derive the Jacobian

Table 1 Properties of null points (type, location, eigenvalues) for the potential, linear force-free fields with $\alpha = [-0.62, 0.74]$ Mm^{-1} and nonlinear force-free fields with $J_{z0} = [-20, -10, 10, 20]$ mA m^{-2}. [a]One real and two complex eigenvalues (only the real parts are reported).

Model	α or J_{z0}	Type	Location (x_0, y_0, z_0)	Eigenvalues $(\lambda_1 = \pm\rho_J, \lambda_2, \lambda_3)$
Potential		+	(64.30, 72.94, 6.90)	$(-\mathbf{0.026}, 0.021, 5.5 \times 10^{-3})$
Linear	-0.62	+	(63.43, 71.2, 5.63)	$(-\mathbf{0.032}, 0.023, 9.28 \times 10^{-3})$
Force-Free		$-$	(68.8, 134.9, 5.12)	$(9.26, -4.8, -4.8) \times 10^{-5}$ [a]
[α]	0.74	+	(65.8, 74.2, 6.78)	$(-\mathbf{0.026}, 0.021, 5.7 \times 10^{-3})$
		+	(6.14, 85.2, 6 \times 10^{-4})	$(-2.68, 2.28, 0.836) \times 10^{-4}$
		$-$	(64.31, 81.63, 17.62)	$(5.8, -4.5, -1.0) \times 10^{-3}$
		$-$	(4.22, 83.12, 4.85)	$(5.83, -3.17, -3.17) \times 10^{-5}$ [a]
		+	(4.58, 82.44, 7.21)	$(-10, 4.86, 4.86) \times 10^{-5}$ [a]
Nonlinear	-20	+	(65.54, 73.57, 6.31)	$(-\mathbf{0.024}, 0.019, 4.86 \times 10^{-3})$
Force-Free		$-$	(63.54, 79.64, 17.23)	$(4.71, -3.78, -0.88) \times 10^{-3}$
[J_{z0}]		+	(15.70, 108.5, 36.98)	$(-4.38, 3.72, 0.35) \times 10^{-5}$
	-10	+	(64.84, 73.39, 6.77)	$(-\mathbf{0.021}, 0.017, 4.55 \times 10^{-3})$
		$-$	(52.33, 89.19, 30.59)	$(4.02, -3.83, -0.21) \times 10^{-4}$
		+	(40.44, 100.7, 36.39)	$(-1.17, 1.13, 0.028) \times 10^{-4}$
	10	+	(63.87, 72.49, 7.12)	$(-\mathbf{0.020}, 0.016, 4.24 \times 10^{-3})$
		$-$	(30.59, 124.7, 0.06)	$(7.27, -6.62, -0.69) \times 10^{-5}$
	20	+	(63.53, 72.10, 7.65)	$(-\mathbf{0.024}, 0.019, 4.9 \times 10^{-3})$
		$-$	(34.29, 114.3, 0.025)	$(7.6, -5.9, -1.9) \times 10^{-4}$
		$-$	(35.82, 112.5, 14.29)	$(6.8, -6.3, -0.66) \times 10^{-4}$

matrix and then to compute the eigenvalues. To describe these properties, we thus introduce a quantity that helps us to classify the nature of the null points, the spectral radius of the Jacobian matrix [ρ_J], as follows:

$$\rho_J = \max_i \left(|\lambda_i|\right) \quad \text{for } i = 1, 2, 3 \tag{13}$$

where the λ_i are the (real or complex) eigenvalues of the Jacobian matrix. For a Jacobian matrix having two complex-conjugate eigenvalues, the spectral radius [ρ_J] is the real eigenvalue for a divergence-free magnetic field.

In Table 1, we summarise the properties of the null points (type, location, eigenvalues, spectral radius) for the potential field, the linear force-free fields with $\alpha = [-0.62, 0.74]$ Mm^{-1}, and the nonlinear force-free fields with $J_{z0} = [-20, -10, 10, 20]$ mA m^{-2}. The eigenvalues are sorted in such way that the spectral radius corresponds to the absolute value of the first eigenvalue. The null point present in the potential configuration is a positive null point with one negative and two positive eigenvalues. This null point [NP0] is also present in the linear and nonlinear force-free configurations with a slight displacement up and down, left and right depending on the sign of the α values. Thus, the null point [NP0] originally created in the potential field is stable for all force-free models. In particular, NP0 is always a positive null point and the spectral radius is almost constant. Therefore we conjecture that the spectral radius gives a good proxy for the stability of a

Figure 9 Magnetic energy $[E_m]$ relative to the potential field energy $[E_{pot}]$ as a function of the force-free parameter $[\alpha: Mm^{-1}]$ for the linear force-free extrapolations. The solid lines indicate the relative energy levels E_m/E_{pot} for the nonlinear force-free extrapolations for $J_{z0} = [-20, -10, 10, 20]$ mA m^{-2}.

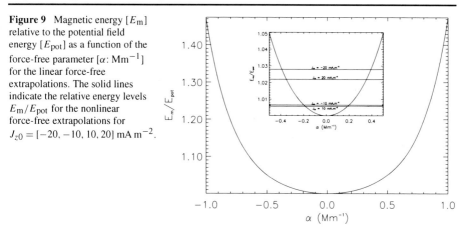

null point in a magnetic-field configuration. Note that a large spectral radius indicates large magnetic-field gradients.

In addition to NP0, other null points can appear in the magnetic configuration depending on the strength of the current density. It is noticeable that null points are mostly created in pairs (negative and positive null points) or they have complex eigenvalues. The latter case, which indeed cannot exist in the force-free assumption, corresponds to null points appearing at locations where

i) the Taylor expansion in the vicinity of the null point is not valid anymore, or

ii) the null-point finder algorithm breaks down, or

iii) the Jacobian matrix elements cannot be derived with enough accuracy (especially in weak-field regions).

Note that we only found four values of α (among 120) for which complex-conjugate eigenvalues exist.

6. Magnetic Energy Budget

6.1. Total Magnetic Energy

In Figure 9, we plot the magnetic energy $[E_m]$ above the potential-field energy $[E_{pot}]$ in the computational volume $[\Omega]$. As the potential field is a minimum-energy state, E_m is always above E_{pot}. The latter inequality is true if and only if both the force-free and potential fields are computed with the same normal component of the magnetic field on each side of the computational box.

The energy curve as a function of α is similar to the second-order polynomial curve obtained in Figure 11 of Régnier and Priest (2007) for a solar active region. For α ranging from -1 to 1 Mm^{-1}, the magnetic energy of linear force-free fields is not more 50% of the potential-field energy. The curve of the free magnetic energy as a function of α is not symmetric with respect to the potential field $[\alpha = 0]$: the magnetic energy is increasing more rapidly for the positive values of α.

The different levels of magnetic energy for the nonlinear force-free fields are plotted in Figure 9 as straight solid lines: $E_m/E_{pot} = [1.0279, 1.0065, 1.0058, 1.022]$ for $J_{z0} = [-20, -10, 10, 20]$ mA m^{-2}, respectively. This small amount of magnetic energy stored in

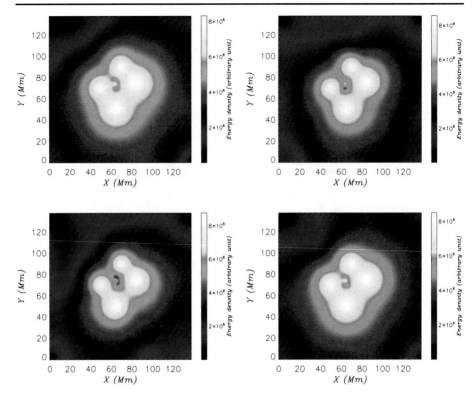

Figure 10 Energy-density maps integrated along the z-axis (in arbitrary unit) for the potential field (top left), the force-free field with $\alpha = -0.62$ Mm^{-1} (top right) and $\alpha = 0.74$ Mm^{-1} (bottom left), and the nonlinear force-free field with $J_{z0} = 20$ mA m^{-2} (bottom right). The colour bar indicates the energy density in arbitrary units.

the nonlinear force-free configurations (less than 3%) is a consequence of the particular current distribution: the ring distribution has no net current in a single polarity and return currents on the edges of the flux bundles, which confine the magnetic field in strong-field regions without generating twisted flux bundles.

6.2. Energy Density Distributions

In Figure 10, we plot the magnetic-energy density integrated along the z-axis to study the distribution of the magnetic energy of the four magnetic fields analysed above. For the potential field, the distribution of the energy density is dominated by the energy (or field strength) near the bottom boundary for the five magnetic polarities. We notice that there is a local minimum of magnetic-energy density where the null point NP0 is located. In addition, there is another obvious local minimum located on the other side of the parasitic polarity with respect to NP0 (location: $x = 62$, $y = 78$). For the two linear force-free fields, the magnetic-energy-density distribution is dominated by the four polarities of the initial quadrupolar field where the magnetic-field strength is large, whilst the parasitic polarity does not influence the distribution. Again we notice that there is a strong local minimum at the location of NP0 and in addition is an annulus-like series of local minima connecting NP0 and NP1 around the parasitic polarity. The other null points located in weak-field regions are not observed on the energy-density maps. For the nonlinear force-free field, the energy-density distribution looks very much like the distribution of the potential field with a maximum of energy density slightly increased.

We notice that we are able to easily identify null points as local minima in the distribution of energy density where strong magnetic-field gradients are observed (large spectral radius).

7. Electric Currents

In Figure 11, we plot the electric-current density integrated along the z-axis for three of the four characteristic magnetic-field computations. We have computed the three components of the current density from the curl of the magnetic field and then plotted the current-density strength (or modulus). The potential field has zero electric current (or only tiny currents due to the errors when the magnetic-field components are differentiated). The distribution of the current density is different from the distributions of the energy density (see Figure 10). The five polarities contribute a large amount to the current-density distribution. As noticed for the energy-density distribution, there exists a local minimum where the null point NP0 is located, and an annulus-like series of local minima exists connecting the null points NP0 and NP1 in the linear force-free fields. For the nonlinear force-free-field configuration, the electric-current-density distribution is similar to the linear force-free distribution with a local minimum at the location of NP0.

Except for the location of null points, it is not obvious where to locate with this method the other topological elements where the current density is supposed to be increased. This shows that, for this configuration, the storage of current density along topological elements is not an efficient mechanism compared to the current density stored in the strong-field regions above the magnetic polarities.

8. Discussion and Conclusions

We investigated the changes in the magnetic-field configurations obtained for different force-free models (potential, linear and nonlinear force-free fields) using the same boundary conditions. We analysed the changes in terms of geometry and connectivity of field lines, magnetic energy, and electric-current distributions. We performed this analysis for a continuous magnetic-field distribution with no symmetry. We imposed an electric-current distribution that we have proven to be stable when a large amount of current is injected into the magnetic configuration (Régnier, 2009). Despite the previous works on this topic, we have here provided an extended analysis of magnetic configurations with a topology that has never been performed before.

For this experiment, the initial configuration corresponds to a potential field with five sources (two large bipoles and one parasitic polarity) having a null point NP0 in the corona. By injecting currents in the magnetic configuration, the geometry and topology of the linear and nonlinear force-free configurations are modified such that:

- The geometry of the field lines is modified similarly to previous results obtained on the same topic: statistically, the field lines are higher and longer when the absolute value of electric current is increased.
- The connectivity of the field lines can be strongly modified near the topological elements (where the connectivity is changed rapidly).

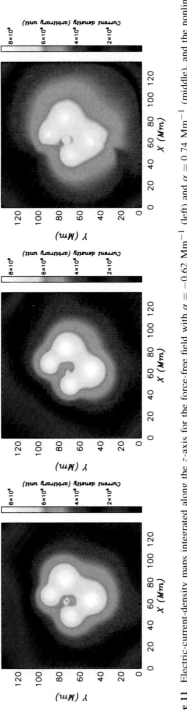
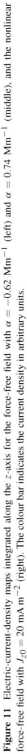

Figure 11 Electric-current-density maps integrated along the z-axis for the force-free field with $\alpha = -0.62$ Mm^{-1} (left) and $\alpha = 0.74$ Mm^{-1} (middle), and the nonlinear force-free field with $J_{z0} = 20$ mA m^{-2} (right). The colour bar indicates the current density in arbitrary units.

- The initial null point NP0 is moved slightly up or down when the force-free parameter [α] varies but remains with the same basic properties, in particular the spectral radius remains almost constant.
- Other null points (up to five) can appear in the magnetic configurations; most of them are located near the boundaries but one [NP1] of which propagates towards the strong-field region when the current density is increased.
- The magnetic energy and current distributions can highlight the location of stable null points where strong magnetic-field gradients are present.

We also noticed that for reasonable values of the electric current injected or the force-free parameter α (< 0.25 Mm^{-1} in this experiment) the magnetic configuration is almost not modified compared to the potential-field configuration.

We thus state that null points existing in potential-field configurations are also present in force-free configurations with the same properties (*e.g.*, sign, spectral radius). This statement means that the null points found in a potential field with a large spectral radius can be considered as stable null points in other magnetic-field models. Even if true for this experiment, we need to confirm this statement in a future statistical study of solar active-region magnetic fields. It is important to note here that this statement is true *i*) when the force-free fields are computed with the same boundary conditions, and *ii*) when there is no noise in the datasets. In Appendix B, we note that the null points in potential-field configurations are slightly affected by boundary conditions (periodic, closed, ...), spatial resolution or size of field-of-view: the null point with the strongest spectral radius remains stable.

From this study, we also provide a benchmark for analysing the topology of a magnetic configuration: it is possible to retrieve important information about the topology just by analysing the distribution of magnetic-energy density and of the electric-current density in the volume. Moreover, we emphasise the importance of checking the divergence-free property of the magnetic field in the vicinity of a null point.

Acknowledgements   SR thanks Eric Priest and Clare Parnell (University of St Andrews) for fruitful discussions on this topic. The computations of force-free field extrapolations have been performed using the XTRAPOL code developed by T. Amari (Ecole Polytechnique, France).

Appendix A: Null Point Description

As a first approximation, we assume that the magnetic field around the null point approaches zero linearly. The magnetic field [**B**] near a neutral point can then be expressed as a first-order Taylor expansion:

$$\mathbf{B} = M \cdot \mathbf{r}, \tag{14}$$

where M is the Jacobian matrix with elements $M_{ij} = \partial B_i / \partial x_j$ for all $i, j = 1, 2, 3$ and $\mathbf{r}$ is the position vector (x, y, z). The Jacobian matrix has some interesting properties:

i) As the magnetic field is divergence-free, $\mathrm{Tr}(M) = 0$ – this property holds at each point of the magnetic-field configuration.

ii) For potential and force-free fields that satisfy $\nabla \wedge \mathbf{B} = 0$ or $\nabla \wedge \mathbf{B} = \alpha \mathbf{B}$, M is symmetric at the location of the null point. Therefore, M has three real eigenvalues and the eigenvectors are orthogonal.

From the first property and for force-free fields, we obtain the following relationship between the three real eigenvalues $(\lambda_1, \lambda_2, \lambda_3)$:

$$\lambda_1 + \lambda_2 + \lambda_3 = 0, \tag{15}$$

meaning that two eigenvalues, λ_2 and λ_3 say, have the same sign, and that $|\lambda_1| = |\lambda_2 + \lambda_3|$ implying that $|\lambda_1| > |\lambda_2|, |\lambda_3|$. According to Parnell *et al.* (1996), the single eigenvalue [λ_1] defines the direction of the spine field line whilst the two other eigenvalues [λ_2, λ_3] indicate the directions of the fan plane. Therefore we can classify null points as positive null points with two positive and one negative eigenvalues, and negative null points with two negative and one positive eigenvalues. For a positive (negative) null point, the magnetic-field lines in the fan plane are directed away from (towards) the null point, whereas the spine field line is pointing towards (away from) the null point. Following Hornig and Schindler (1996), a null point is unstable if $\det(M) = 0$. The determinant of M is

$$\det(M) = \prod_{i=1}^{3} \lambda_i, \tag{16}$$

with $\det(M) < 0$ for a positive null point and $\det(M) > 0$ for a negative one. For potential and force-free fields, null points are stable, except if one eigenvalue [λ_2 or λ_3] vanishes reducing the null point to a 2D null point unstable in the 3D configuration. If λ_1 vanishes, then all eigenvalues have to be zero, and thus the first-order Taylor expansion is no longer valid and the null-point properties are thus derived from the Hessian matrix instead of the Jacobian matrix.

Appendix B: On the Topology of Potential Fields

In this article, we have studied in detail the topology of force-free fields for a distribution of photospheric magnetic field inducing the existence of a null point. In order to compare the magnetic energy and properties of the different configurations, we have used the same boundary conditions. But is a magnetic configuration modified when the boundary conditions are changed? and is the topology influenced by the spatial scales of the bounded box? To address these questions, we perform a comparison of different potential fields and different spatial scales.

B.1. Effect of Boundary Conditions

The above computation was performed using a Grad–Rubin numerical scheme and assuming closed boundary conditions on the sides and top of the computational box. We now compute the magnetic configurations for a potential field with open boundary conditions and with periodic boundary conditions. In Figure 12 top row, we plot the location of the null points for these new boundary conditions. This has to be compared to Figure 6. The location of NP0 is similar for all models and the spectral radius of NP0 is again the strongest (see Table 2). The potential field with periodic boundary conditions has created two negative null points near the sides of the computational box. The topology of the potential field is slightly influenced by the boundary conditions for this experiment.

B.2. Effect of Spatial Resolution

To study the effects of the spatial resolution on the topology of potential fields, we modify the pixel size by a factor of 0.5 and 2 (Figure 12 middle row). We also make sure that the total unsigned magnetic flux remains unchanged (variation less than 1%). The computations are carried out with periodic boundary conditions. NP0 is located nearly at the same location for

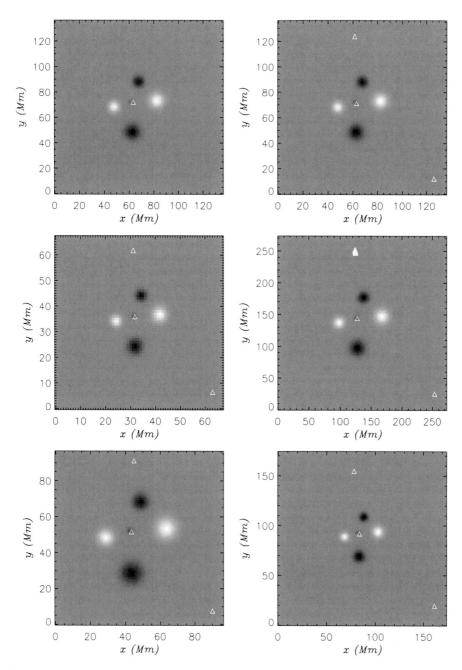

Figure 12 Location of null points (triangles) on the x–y-plane for different potential fields. Top row: potential fields with different boundary conditions: (Left) open sides and top boundaries, (Right) periodic. Middle row: potential fields with a different spatial resolution for the same field of view: (Left) decreased by a factor of two, (Right) increased by a factor of two. Bottom row: potential fields for different field-of-view: (Left) reduced by 20 pixels on each edge, (Right) increased by 20 pixels on each edge.

Table 2 Properties of null points (type, location, eigenvalues) for the different potential fields.

Model	Type	Location $[x_0, y_0, z_0]$	Eigenvalues $[\lambda_1 = \pm \rho_J, \lambda_2, \lambda_3]$
Boundary Conditions			
Closed Boundaries	+	(64.30, 72.94, 6.90)	$(-\mathbf{0.026}, 0.021, 5.5 \times 10^{-3})$
Open Boundaries	+	(63.3, 71.96, 7.39)	$(-\mathbf{0.019}, 0.016, 5.2 \times 10^{-3})$
Periodic Boundaries	+	(62.93, 71.25, 7.91)	$(-\mathbf{0.023}, 0.017, 5.5 \times 10^{-3})$
	−	(61.4, 123.8, 0.02)	$(8.6, -5.1, -3.5) \times 10^{-3}$
	−	(126, 12.15, 5.38)	$(2.3, 0.02, 2.28) \times 10^{-4}$
Spatial Resolution			
$70 \times 70 \times 60$	+	(31.7, 35.8, 3.91)	$(-\mathbf{0.047}, 0.035, 0.01)$
	−	(31.06, 61.75, 0.01)	$(1.9, -1.1, -0.7) \times 10^{-3}$
	−	(63.05, 6.51, 3.14)	$(4.73, -4.68, -0.04) \times 10^{-4}$
$280 \times 280 \times 240$	+	(127.37, 144, 15.84)	$(-\mathbf{0.011}, 2.8 \times 10^{-3}, 8.7 \times 10^{-3})$
	−	(124.74, 246.8, 0.022)	$(4.7, -2.76, -1.92) \times 10^{-4}$
	−	(124.66, 247.22, 0.022)	$(4.6, -2.72, -1.9) \times 10^{-4}$
	−	(124.4, 248.57, 0.023)	$(4.4, -2.58, -1.81) \times 10^{-4}$
	−	(124.22, 249.57, 0.024)	$(4.25, -2.5, -1.75) \times 10^{-4}$
	−	(124.08, 250.32, 0.024)	$(4.15, -2.44, -1.71) \times 10^{-4}$
	−	(123.8, 251.91, 0.025)	$(3.97, -2.32, -1.64) \times 10^{-4}$
	−	(123.77, 252.06, 0.025)	$(3.96, -2.31, -1.63) \times 10^{-4}$
	−	(253.49, 25.76, 8.39)	$(1.15, -1.16, -0.007) \times 10^{-4}$
Field-of-View			
100×100	+	(43.5, 51.59, 7.83)	$(-\mathbf{0.024}, 0.017, 5.7 \times 10^{-3})$
	−	(44.9, 91.1, 0.18)	$(5.5, -2.9, -2.6) \times 10^{-3}$
	−	(90.07, 7.59, 14.8)	$(8.8, -8.4, -0.48) \times 10^{-4}$
180×180	+	(83.41, 91.8, 7.92)	$(-\mathbf{0.023}, 0.017, 5.5 \times 10^{-3})$
	−	(77.9, 154.57, 0.025)	$(2.6, -1.64, -0.99) \times 10^{-4}$
	−	(161.7, 19.45, 1.16)	$(7.93, -7.72, -2.18) \times 10^{-5}$

both spatial resolutions. The other null points are similar to those appearing in the potential field with periodic conditions. The cluster of eight null points suggests that there is a null line at this location. The topology of the potential field is thus similar to the topology of the potential field with closed boundary conditions. We again emphasise that the spectral radius of NP0 remains almost constant and is the strongest (see Table 2).

B.3. Effect of Field-of-View

We now modify the field-of-view of the experiment but keep the same spatial resolution and the same total unsigned magnetic flux. In Figure 12 bottom row, we plot the location of the null points for two different fields-of-view using a potential field with periodic conditions: we decrease or increase the number of pixels by 40 pixels in each direction (20 pixels on each edge). The properties of the null points are summarised in Table 2. The number of null points is the same for the different fields-of-view. The spectral radius of NP0 is similar and still the strongest. The topology of the magnetic configuration is almost not influenced by the field-of-view for this experiment. It is worth noticing that the field-of-view was modified in such way that there is no magnetic flux on the edges of the computational box as this would strongly influence the magnetic configuration and also that it would have unexpected effects due to the violation of the solenoidal condition.

References

Amari, T., Boulmezaoud, T.Z., Aly, J.J.: 2006, *Astron. Astrophys.* **446**, 691 – 705. doi:10.1051/0004-6361: 20054076.

Amari, T., Aly, J.J., Luciani, J.F., Boulmezaoud, T.Z., Mikic, Z.: 1997, *Solar Phys.* **174**, 129 – 149.

Amari, T., Boulmezaoud, T.Z., Mikic, Z.: 1999, *Astron. Astrophys.* **350**, 1051.

Antiochos, S.K., Devore, C.R., Klimchuk, J.A.: 1999, *Astrophys. J.* **510**, 485 – 493.

Aulanier, G., DeLuca, E.E., Antiochos, S.K., McMullen, R.A., Golub, L.: 2000, *Astrophys. J.* **540**, 1126 – 1142. doi:10.1086/309376.

Barnes, G.: 2007, *Astrophys. J.* **670**, 53 – 56. doi:10.1086/524107.

Barnes, G., Longcope, D.W., Leka, K.D.: 2005, *Astrophys. J.* **629**, 561 – 571. doi:10.1086/431175.

Berger, M.A., Prior, C.: 2006, *J. Phys. A, Math. Gen.* **39**, 8321 – 8348. doi:10.1088/0305-4470/39/26/005.

Brown, D.S., Priest, E.R.: 1999, *Proc. Roy. Soc. London Ser. A, Math. Phys. Sci.* **455**, 3931 – 3951.

Brown, D.S., Priest, E.R.: 2000, *Solar Phys.* **194**, 197 – 204.

Bungey, T.N., Titov, V.S., Priest, E.R.: 1996, *Astron. Astrophys.* **308**, 233 – 247.

Close, R.M., Heyvaerts, J.F., Priest, E.R.: 2004, *Solar Phys.* **225**, 267 – 292. doi:10.1007/s11207-004-4279-5.

Démoulin, P., Hénoux, J.C., Mandrini, C.H.: 1994, *Astron. Astrophys.* **285**, 1023 – 1037.

Deng, N., Liu, C., Yang, G., Wang, H., Denker, C.: 2005, *Astrophys. J.* **623**, 1195 – 1201. doi:10.1086/428821.

Grad, H., Rubin, H.: 1958, In: *Proc. 2nd Int. Conf. on Peaceful Uses of Atomic Energy* **31**, UN, Geneva, 190.

Haynes, A.L., Parnell, C.E.: 2007, *Phys. Plasmas* **14**, 2107. doi:10.1063/1.2756751.

Hornig, G., Schindler, K.: 1996, *Phys. Plasmas* **3**, 781 – 791.

Hudson, T.S., Wheatland, M.S.: 1999, *Solar Phys.* **186**, 301 – 310.

Janse, Å.M., Low, B.C., Parker, E.N.: 2010, *Phys. Plasmas* **17**(9), 092901. doi:10.1063/1.3474943.

Leka, K.D., Skumanich, A.: 1999, *Solar Phys.* **188**, 3 – 19.

Li, H., Schmieder, B., Aulanier, G., Berlicki, A.: 2006, *Solar Phys.* **237**, 85 – 100. doi:10.1007/s11207-006-0173-7.

Low, B.C.: 2006, *Astrophys. J.* **649**, 1064 – 1077. doi:10.1086/506586.

Low, B.C.: 2007, *Phys. Plasmas* **14**(12), 122904. doi:10.1063/1.2822151.

Luoni, M.L., Mandrini, C.H., Cristiani, G.D., Démoulin, P.: 2007, *Adv. Space Res.* **39**, 1382 – 1388. doi:10.1016/j.asr.2007.02.005.

Magara, T., Longcope, D.W.: 2003, *Astrophys. J.* **586**, 630 – 649.

Parnell, C.E., Smith, J.M., Neukirch, T., Priest, E.R.: 1996, *Phys. Plasmas* **3**, 759 – 770.

Priest, E.R., Forbes, T.: 2000, *Magnetic Reconnection: MHD Theory and Applications*, Cambridge University Press, Cambridge, 34 – 37.

Priest, E.R., Longcope, D.W., Heyvaerts, J.: 2005, *Astrophys. J.* **624**, 1057 – 1071. doi:10.1086/429312.

Régnier, S.: 2007, *Mem. Soc. Astron. Ital.* **78**, 126.

Régnier, S.: 2009, *Astron. Astrophys.* **497**, 17 – 20. doi:10.1051/0004-6361/200811502.

Régnier, S., Canfield, R.C.: 2006, *Astron. Astrophys.* **451**, 319 – 330.

Régnier, S., Priest, E.R.: 2007, *Astron. Astrophys.* **468**, 701 – 709.

Régnier, S., Parnell, C.E., Haynes, A.L.: 2008, *Astron. Astrophys.* **484**, 47 – 50. doi:10.1051/0004-6361: 200809826.

Santos, J.C., Büchner, J., Otto, A.: 2011, *Astron. Astrophys.* **525**, 3. doi:10.1051/0004-6361/201014758.

Schrijver, C.J., Title, A.M.: 2002, *Solar Phys.* **207**, 223 – 240.

Subramanian, S., Madjarska, M.S., Maclean, R.C., Doyle, J.G., Bewsher, D.: 2008, *Astron. Astrophys.* **488**, 323 – 329. doi:10.1051/0004-6361:20079315.

Wiegelmann, T.: 2008, *J. Geophys. Res.* **113**(A12), 3. doi:10.1029/2007JA012432.

Zhao, H., Wang, J.X., Zhang, J., Xiao, C.J., Wang, H.M.: 2008, *Chin. J. Astron. Astrophys.* **8**, 133 – 145. doi:10.1088/1009-9271/8/2/01.

Solar Phys (2012) 277:153–163
DOI 10.1007/s11207-011-9816-4

Can We Determine Electric Fields and Poynting Fluxes from Vector Magnetograms and Doppler Measurements?

G.H. Fisher · B.T. Welsch · W.P. Abbett

Received: 21 January 2011 / Accepted: 9 June 2011 / Published online: 26 July 2011
© Springer Science+Business Media B.V. 2011

Abstract The availability of vector-magnetogram sequences with sufficient accuracy and cadence to estimate the temporal derivative of the magnetic field allows us to use Faraday's law to find an approximate solution for the electric field in the photosphere, using a Poloidal–Toroidal Decomposition (PTD) of the magnetic field and its partial time derivative. Without additional information, however, the electric field found from this technique is under-determined – Faraday's law provides no information about the electric field that can be derived from the gradient of a scalar potential. Here, we show how additional information in the form of line-of-sight Doppler-flow measurements, and motions transverse to the line-of-sight determined with *ad-hoc* methods such as local correlation tracking, can be combined with the PTD solutions to provide much more accurate solutions for the solar electric field, and therefore the Poynting flux of electromagnetic energy in the solar photosphere. Reliable, accurate maps of the Poynting flux are essential for quantitative studies of the buildup of magnetic energy before flares and coronal mass ejections.

Keywords Flares, dynamics · Helicity, magnetic · Magnetic fields, corona

1. Introduction

The launch of SDO, with its ability to measure the Sun's vector magnetic field anywhere on the disk with a high temporal cadence, promises to usher in a new era of solar astronomy. This new era of measurement demands new approaches for the analysis and use of these

Solar Flare Magnetic Fields and Plasmas
Guest Editors: Y. Fan and G.H. Fisher

G.H. Fisher (✉) · B.T. Welsch · W.P. Abbett
Space Sciences Laboratory, University of California, Berkeley, CA, USA
e-mail: fisher@ssl.berkeley.edu

B.T. Welsch
e-mail: welsch@ssl.berkeley.edu

W.P. Abbett
e-mail: abbett@ssl.berkeley.edu

data. We show in this article how the vector magnetic field and Doppler-flow measurements that can now be made with HMI (Scherrer and The HMI Team, 2005) lead to new methods for determining the electric field vector and the Poynting-flux vector

$$\mathbf{S} = \frac{1}{4\pi} c\mathbf{E} \times \mathbf{B} \tag{1}$$

at the solar photosphere. The Poynting flux measures the flow of electromagnetic energy at the layers where the magnetic field is determined. Quantitative observational studies of how energy flows into the corona depend on deriving accurate estimates of the Poynting flux.

Most work estimating the Sun's electric field or Poynting flux either explicitly or implicitly assumes that the electric field is determined by ideal MHD processes, and therefore the problem can be reduced to determining a velocity field associated with the observed magnetic-field evolution. One class of velocity-estimation techniques are known as "Local Correlation Tracking" (LCT) methods, which essentially capture pattern motions of the line-of-sight magnetic field or white-light intensity. This approach was pioneered by November and Simon (1988). Other implementations include the Lockheed–Martin LCT code (Title *et al.*, 1995; Hurlburt *et al.*, 1995), "Balltracking" (Potts, Barrett, and Diver, 2004), and the FLCT code (Fisher and Welsch, 2008). Another class of velocity-estimation methods incorporate solutions of the vertical component of the magnetic induction equation into determinations of the velocity field (Kusano *et al.*, 2002; Welsch *et al.*, 2004; Longcope, 2004; Schuck, 2006, 2008; Chae and Sakurai, 2008). The work that we present in this article incorporates solutions of the three-dimensional magnetic-induction equation, using the electric field as the fundamental variable, rather than the velocity field.

The temporal evolution of the Sun's magnetic field is governed by Faraday's law,

$$\frac{\partial \mathbf{B}}{\partial t} = -\nabla \times c\mathbf{E}. \tag{2}$$

If one can make a map on the photosphere of $\partial\mathbf{B}/\partial t$, can one determine $\mathbf{E}$ by uncurling this equation? Addressing this question was the focus of Fisher *et al.* (2010), in which a poloidal–toroidal decomposition (PTD) of the temporal derivative of the magnetic field was used to invert Faraday's law to find $\mathbf{E}$. Fisher *et al.* (2010) found that one could indeed find solutions for $\mathbf{E}$ that solve all three components of Faraday's law, but the solutions are not unique: the gradient of a scalar function can be added to the PTD solutions for $\mathbf{E}$ without affecting $\nabla \times \mathbf{E}$. Fisher *et al.* (2010) explored two different methods for determining the scalar function using *ad-hoc* and variational methods, both of which enforced the assumption, from ideal MHD, that $\mathbf{E}$ must be normal to $\mathbf{B}$. Unfortunately, the agreement with a test case from an MHD solution, while better than conventional correlation-tracking methods, was still disappointing. The authors concluded that including additional information from other observed data was one possible approach for improving the electric field inversions.

In this article, we use the same MHD simulation test case used in Welsch *et al.* (2007) and Fisher *et al.* (2010) to show that using Doppler-flow measurements to determine the electric scalar potential, especially in regions where the magnetic field is primarily horizontal, can dramatically improve the inversion for the electric field and the Poynting flux.

In Section 2 we review the PTD formalism, which describes how one can derive the purely inductive part of the electric field from measurements that estimate the time derivative of $\mathbf{B}$, and the technique of Section 3.2 of Fisher *et al.* (2010), showing how one can derive a potential electric field, which, when added to the inductive part of the electric field, is normal to the magnetic field. This is useful in generating electric-field solutions that are both consistent with Faraday's law and with ideal MHD, which is generally believed to be a good approximation in the solar photosphere.

Section 3 argues from physical grounds why magnetic-flux emergence may make a large contribution to the part of the electric field attributable to a potential function. Then, starting from this argument, we derive a Poisson equation for an electric-field potential function that is determined primarily from knowledge of the vertical velocity field, as determined from Doppler measurements, and the horizontal magnetic field near polarity inversion lines where the field is nearly horizontal. The electric field from this contribution is then added to that determined from the PTD solutions. We then apply this technique to the MHD simulation test data, to compare the electric field from the simulation with that from PTD alone, and with that from combining PTD with Doppler measurements.

In Section 4, we try a similar approach, but instead of using contributions to the horizontal electric field from Doppler measurements, we use non-inductive contributions to the electric field determined from the FLCT correlation-tracking technique, applicable in regions where the magnetic field is mainly vertical. This technique is essentially the three-dimensional analogue of the ILCT technique described by Welsch *et al.* (2004). We also try combining PTD with contributions from both the Doppler measurements and those from FLCT, and compare with the simulation data.

Our results are summarized in Section 5, along with a discussion of where additional work is needed.

2. Poloidal–Toroidal Decomposition

Here, we present only a brief synopsis of the PTD method of deriving an electric field [**E**] that obeys Faraday's law. More details can be found in Section 2 of Fisher *et al.* (2010).

Since the three-dimensional magnetic-field vector is a solenoidal quantity, one can express the magnetic field in terms of two scalar functions [$\mathcal{B}$ (the "poloidal" potential) and $\mathcal{J}$ (the "toroidal" potential)] as follows:

$$\mathbf{B} = \nabla \times \nabla \times \mathcal{B}\hat{\mathbf{z}} + \nabla \times \mathcal{J}\hat{\mathbf{z}}. \tag{3}$$

Taking the partial time derivative of Equation (3) one finds

$$\dot{\mathbf{B}} = \nabla \times \nabla \times \dot{\mathcal{B}}\hat{\mathbf{z}} + \nabla \times \dot{\mathcal{J}}\hat{\mathbf{z}}. \tag{4}$$

Here, the overdot denotes a partial time derivative. We will now assume a locally Cartesian coordinate system, in which the directions parallel to the photosphere are denoted with a "horizontal" subscript h, and the vertical direction is denoted with subscript z. One can then re-write Equations (3) and (4) in terms of horizontal and vertical derivatives as

$$\mathbf{B} = \nabla_h \left(\frac{\partial \mathcal{B}}{\partial z} \right) + \nabla_h \times \mathcal{J}\hat{\mathbf{z}} - \nabla_h^2 \mathcal{B}\hat{\mathbf{z}}, \tag{5}$$

and

$$\dot{\mathbf{B}} = \nabla_h \left(\frac{\partial \dot{\mathcal{B}}}{\partial z} \right) + \nabla_h \times \dot{\mathcal{J}}\hat{\mathbf{z}} - \nabla_h^2 \dot{\mathcal{B}}\hat{\mathbf{z}}. \tag{6}$$

One useful property of the poloidal–toroidal decomposition is that the scalar functions $\dot{\mathcal{B}}$, $\dot{\mathcal{J}}$, and $\partial\dot{\mathcal{B}}/\partial z$ can all be determined by knowing the time derivative of the magnetic-field vector in the plane of the photosphere. By examining the z-component of Equation (6), the z-component of the curl of Equation (6), and the horizontal divergence of Equation (6), one can derive the following three two-dimensional Poisson equations for $\dot{\mathcal{B}}$, $\dot{\mathcal{J}}$, and $\partial\dot{\mathcal{B}}/\partial z$:

$$\nabla_h^2 \dot{\mathcal{B}} = -\dot{B}_z, \tag{7}$$

$$\nabla_h^2 \dot{\mathcal{J}} = -(4\pi/c)\dot{J}_z = -\hat{\mathbf{z}} \cdot \left(\nabla \times \dot{\mathbf{B}}_h\right), \tag{8}$$

and

$$\nabla_h^2\left(\partial\dot{\mathcal{B}}/\partial z\right) = \nabla_h \cdot \dot{\mathbf{B}}_h. \tag{9}$$

Here, $\dot{B}_z$ and $\dot{\mathbf{B}}_h$ denote the partial time derivatives of the vertical and horizontal components of the magnetic field, respectively. Solving these three Poisson equations provides sufficient information to determine an electric field that satisfies Faraday's law.

By comparing the form of Equation (2) with Equations (4) and (6) it is clear that the following must be true:

$$\nabla \times c\mathbf{E} = -\nabla \times \nabla \times \dot{\mathcal{B}}\hat{\mathbf{z}} - \nabla \times \dot{\mathcal{J}}\hat{\mathbf{z}} \tag{10}$$

$$= -\nabla_h(\partial\dot{\mathcal{B}}/\partial z) - \nabla_h \times \dot{\mathcal{J}}\hat{\mathbf{z}} + \nabla_h^2\dot{\mathcal{B}}\hat{\mathbf{z}}. \tag{11}$$

Uncurling Equation (10) yields this expression for the electric field $\mathbf{E}$:

$$c\mathbf{E} = -\nabla \times \dot{\mathcal{B}}\hat{\mathbf{z}} - \dot{\mathcal{J}}\hat{\mathbf{z}} - c\nabla\psi \equiv c\mathbf{E}^I - c\nabla\psi. \tag{12}$$

Here, $-\nabla\psi$ is the contribution to the electric field from a scalar potential, for which solutions to Faraday's law reveal no information. The solution for $\mathbf{E}$ without the contribution from $-\nabla\psi$ [$\mathbf{E}^I$] is the purely inductive solution determined from the PTD method. Within this article, this solution will be referred to simply as the PTD solution or the PTD electric field. Note that the PTD solution is not unique. While solutions for $\partial\dot{\mathcal{B}}/\partial z$ are necessary to ensure that Faraday's law is obeyed, the PTD solution for the electric field itself depends only on $\dot{\mathcal{B}}$ and $\dot{\mathcal{J}}$. This means that the PTD electric field is the same for distributions of $\dot{B}_z$ and $\dot{\mathbf{B}}_h$ which have differing values of $\nabla_h \cdot \dot{\mathbf{B}}_h$, but the same values of $(\nabla_h \times \dot{\mathbf{B}}_h) \cdot \hat{\mathbf{z}}$ and $\dot{B}_z$. Thus the PTD solutions for $\mathbf{E}^I$ are under-determined.

Fisher *et al.* (2010) described two techniques for deriving an electric-field contribution from a scalar potential, in an effort to resolve the under-determined nature of the PTD solutions. The first technique, described in Section 3.2 of that article, presents an *ad-hoc* iterative method for deriving a scalar potential electric field that, when added to the PTD solution, results in an electric field that is normal to $\mathbf{B}$, and hence consistent with ideal MHD. The second technique, based on a variational method, finds a scalar potential electric field that, when added to the PTD solution, minimizes the area integral of $|\mathbf{E}|^2$ or $|\mathbf{v}|^2$. When compared to the original electric field from the simulation test case, the iterative method applied to the PTD solutions showed a qualitative consistency, but not detailed agreement with the simulation electric field, while the electric field computed with the variational technique showed poor agreement. Fisher *et al.* (2010) concluded that significant improvement in the agreement of the inverted electric field with the real electric field requires additional observational information beyond the temporal evolution of $\mathbf{B}$.

3. The Importance of Doppler Flow Measurements to the Electric Field

We argue here that when flux emergence occurs, much of the missing information about non-inductive contributions to the electric field is contained in Doppler-flow information (see also Ravindra, Longcope, and Abbett, 2008 and Schuck *et al.*, 2010), particularly near polarity inversion lines (PILs), where the horizontal magnetic field is much stronger than the vertical field. We illustrate this point with a simple thought-experiment, shown schematically in Figure 1. Consider the emergence of new magnetic flux in an idealized bipolar-flux system, where the PIL maintains its orientation as flux continuously emerges from below the photosphere. Imagine that vector-magnetogram and Doppler observations are taken from a vantage point normal to the solar surface. Let us focus attention on what is happening near the center of the PIL. Suppose the magnetic field there remains time-invariant as flux continues to emerge, so the time derivative of the magnetic field there is zero, implying that Faraday's law cannot be used to infer the physics of the emerging flux. Yet the electric field at this location should be very large, driven by the upward motion of the plasma carrying the strong, horizontal field. In this case, magnetic-flux emergence will have a strong inductive signature at the edges of the idealized active region, where the vertical magnetic field is changing rapidly, but not near the center of the PIL. Thus, it seems plausible that the electric field near PILs in more realistic emerging-flux configurations will have a significant non-inductive component.

Starting from this perspective, we have explored enhancements to the PTD method that use Doppler-flow information to more tightly constrain the PTD electric-field solutions, with the additional assumption that the photospheric electric field is primarily governed by ideal MHD processes. Directly above PILs, the vertical velocity and the observed horizontal component of the magnetic field unambiguously determine the horizontal electric field:

$$c\mathbf{E}_h^D = -v_z\hat{\mathbf{z}} \times \mathbf{B}_h, \tag{13}$$

where we assume that $|B_z|/|B_h|$ is small. If we can use line-of-sight Doppler-velocity measurements to estimate v_z, we add a powerful constraint to the PTD solution for the electric

Figure 1 Schematic illustration of the emergence of new flux over a time interval Δt, viewed in a vertical plane normal to the polarity inversion line (PIL) in an idealized bipolar flux system. The emerging flux is rising at a speed v_z, which could be inferred by the Doppler shift measured by an observer viewing the PIL from above. The width of the bipolar flux system (the distance from the outer edge of one pole to the outer edge of the other pole) at the beginning of Δt is $2x_0$. Notice that the change in B_z at the outer edges of the emerging-flux region is large, while the change in B_z at the PIL itself – where the flux is actually emerging – is zero (see text).

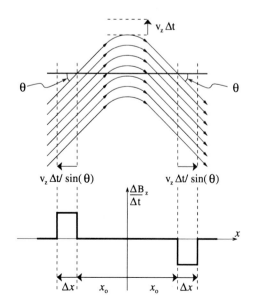

field. Of course, we would like to use the Doppler information away from PILs as well, but are hindered by two complications: *i*) flows parallel to the magnetic field will not affect the electric field at all, but may contribute to the observed Doppler velocity signal, and *ii*) when the vertical component of the magnetic field $[B_z]$ becomes significant compared to the horizontal field $[B_h]$, there is an additional contribution to the horizontal electric field from flow parallel to the surface, which is not accounted for.

We now develop a formal solution for a non-inductive contribution to the electric field that includes information from Doppler-shift measurements, and apply it to a test case with a known electric field. First, from the pair of synthetic vector magnetograms taken from the ANMHD simulation test case described in Welsch *et al.* (2007) and Section 3.1 of Fisher *et al.* (2010), we use the PTD method to find an electric-field solution, neglecting any contribution from a scalar electric-field potential function. We use the numerical techniques and boundary conditions described in Section 3.1 of Fisher *et al.* (2010). Second, we compute a candidate horizontal electric field from vertical velocities taken from the simulation as synthetic Doppler-flow measurements, and horizontal magnetic fields from the synthetic vector magnetograms, from Equation (13) above. This electric field is then multiplied by a "confidence function", which is near unity at PILs, but decreases to zero when $|B_z/B_h|$ is no longer small. This reflects our lack of confidence in the accuracy of this horizontal Doppler electric field in those locations, for the reasons described earlier. The specific form for the confidence function is probably not important. Here, we assume that the confidence function w is given by

$$w = \exp\left[-\left(|B_z|/|B_h|\right)^2/\sigma^2\right],\qquad(14)$$

where σ is a free parameter that can be adjusted, and in the specific cases shown in this article was set somewhat arbitrarily to 0.6. We define the "modulated" electric field within the plane of the magnetogram as

$$\mathbf{E}_h^M = w\mathbf{E}_h^D.\qquad(15)$$

Third, we take the divergence of this modulated horizontal electric field $\mathbf{E}_h^M$, and find the electric-potential function that can best represent it by setting

$$c\mathbf{E}^\chi = -\nabla_h\chi,\qquad(16)$$

where χ solves the Poisson equation

$$\nabla_h^2\chi = -\nabla_h\cdot c\mathbf{E}_h^M.\qquad(17)$$

Because the synthetic vector magnetograms and Doppler flows taken from the MHD simulation use periodic boundary conditions, we use FFT techniques to solve Equation (17). Adding this contribution onto the PTD solutions means that information about the electric field at PILs has been incorporated, while also maintaining consistency with Faraday's Law, since $\mathbf{E}^\chi$ has no curl. Since we generally expect ideal MHD to be a good approximation for conditions in the solar photosphere, we next remove the components of $\mathbf{E}$ parallel to $\mathbf{B}$ by adding the electric field from a second potential function $[\psi]$, using the iterative technique described in Section 3.2 of Fisher *et al.* (2010):

$$c\mathbf{E}^{\text{tot}} = c\mathbf{E}^I + c\mathbf{E}^\chi - \nabla\psi,\qquad(18)$$

where $\nabla\psi\cdot\mathbf{B} = (c\mathbf{E}^I + c\mathbf{E}^\chi)\cdot\mathbf{B}$.

The resulting solutions for $\mathbf{E}$ are shown in the third row of Figure 2, with a scatterplot comparison of S_z of the PTD method and the PTD plus Doppler information with the actual simulation electric fields shown in the top two panels of Figure 3. These portions of the figures show that the recovery of the electric-field components and the Poynting flux is dramatically better than PTD alone.

4. How Important Are Horizontal, Non-inductive Flows?

In the previous section, we considered the role of Doppler-flow measurements in determining non-inductive contributions to the horizontal electric field, and found that combining this information with the PTD solutions for Faraday's law results in a dramatic improvement in the recovery of the electric field. However, this treatment neglects possible contributions to the horizontal electric field away from PILs where a cross product of horizontal velocity with vertical magnetic field could also contribute to the horizontal electric field. Contributions to the horizontal electric field that solve the induction equation have already been incorporated by the PTD solutions, but as with vertical velocities, there could be a sub-space of horizontal flows that do not contribute to Faraday's law.

To evaluate this effect, we estimate horizontal velocities using the FLCT local-correlation tracking (LCT) code (see Fisher and Welsch, 2008), available from http://solarmuri.ssl. berkeley.edu/~fisher/public/software/FLCT/C_VERSIONS/ using images of the vertical component of the magnetic field. Velocities were not computed for pixels with a vertical magnetic field strength below 370 G (see discussion in Welsch *et al.*, 2007), with the windowing parameter σ set to five pixels. The low-pass filtering option was not invoked. The result is a map of the apparent horizontal-velocity field [$\mathbf{U}_h \equiv U_x \hat{\mathbf{x}} + U_y \hat{\mathbf{y}}$] computed at the strong vertical magnetic-field locations, and with velocities at all other locations set to zero. A candidate horizontal electric field is estimated by setting

$$c\mathbf{E}_h^{\mathrm{LCT}} = -\mathbf{U}_h \times \hat{\mathbf{z}} B_z. \tag{19}$$

To consider only non-inductive contributions from $\mathbf{U}_h$, we perform the same general operation as in the previous section, namely to multiply $\mathbf{E}_h^{\mathrm{LCT}}$ by a confidence function, and then eliminate the inductive part of the electric field. Here, the confidence function will be the complement of the confidence function used for the Doppler case, since the LCT estimates are nearly useless near PILs, where the Doppler results should be reliable, while the LCT results should be best when the magnetic field is mostly vertical (and where the Doppler measurements are useless).

We define $c\mathbf{E}^\zeta = -\nabla_h \zeta$, and assume that

$$\nabla_h^2 \zeta = -\nabla_h \cdot (1 - w) c\mathbf{E}_h^{\mathrm{LCT}}. \tag{20}$$

Once this equation has been solved and $\mathbf{E}^\zeta$ has been computed, it can be added to the PTD solutions for $\mathbf{E}^I$, and as in the previous section, a second potential solution can be found that eliminates components of $\mathbf{E}$ parallel to $\mathbf{B}$. Note that combining the PTD solutions with $\mathbf{E}^\zeta$ in this way is like the approach used in the ILCT technique described by Welsch *et al.* (2004), except that solutions of a single component of the induction equation are replaced by solutions to all three components of the induction equation.

The resulting electric field and Poynting flux can be compared to the actual case, the un-altered PTD case, and the case where only the Doppler information is used. The electric field and Poynting-flux results are shown as the fourth row of panels in Figure 2 and a

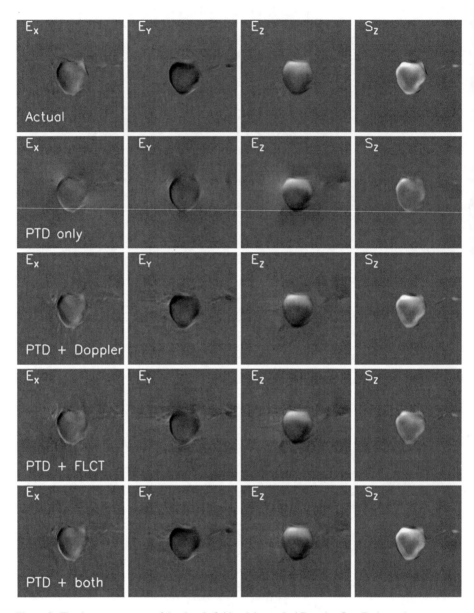

Figure 2 The three components of the electric field and the vertical Poynting flux. Each panel represents an area approximately 100 Mm on a side. Top row: The MHD reference simulation of emerging magnetic flux in a turbulent convection zone. See Welsch *et al.* (2007) for details. Second row: The inductive components of **E** and S_z determined using the PTD method. Third row: **E** and S_z derived by incorporating Doppler flows around PILs into the PTD solutions. Note the dramatic improvement in the estimate of S_z. Fourth row: **E** and S_z derived by incorporating only non-inductive FLCT-derived flows into the PTD solutions. Note the poorer recovery of E_x, E_y, and S_z relative to the case that included only Doppler flows. Fifth row: **E** and S_z derived by including both Doppler flows and non-inductive FLCT flows into the PTD solutions. Note the good recovery of E_x, E_y, and S_z, and the reduction in artifacts in the low-field regions for E_y (best viewed in the electronic version of the article).

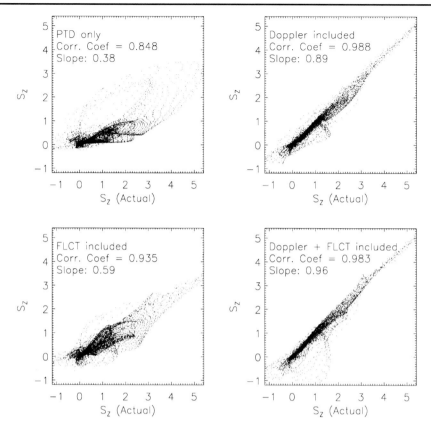

Figure 3 Upper left: A comparison of the vertical component of the Poynting flux derived from the PTD method alone with the actual Poynting flux of the MHD reference simulation. Upper right: A comparison between the simulated results and the improved technique that incorporates information about the vertical flow field around PILs into the PTD solutions. Lower left: Comparison of the vertical Poynting flux when non-inductive FLCT-derived flows are incorporated into the PTD solutions. Lower right: Comparison of the vertical Poynting flux when both Doppler-flow information and non-inductive FLCT-derived flows are incorporated into the PTD solutions. Each scatterplot also shows the computed linear correlation coefficient, as well as the slope of the fit derived with IDL's LADFIT function. Poynting-flux units are in $[10^5 \, \mathrm{G}^2 \, \mathrm{km \, s}^{-1}]$.

scatterplot of the Poynting-flux values with the actual values is shown in the lower-left panel of Figure 3. While the overall performance of the FLCT case is better than that of PTD alone, it is not significantly better than simply applying the iterative method directly to the PTD results as was described in Section 3.2 of Fisher *et al.* (2010). It is definitely not as good as the performance that we show from the Doppler-only case. We conclude that most of the useful information about the non-inductive electric field, at least for this particular simulation of strong flux emergence, is contained within the Doppler flow information.

Does the LCT information, when added to the Doppler-flow information, significantly improve the resulting estimate for the electric field? To answer this question, we have added both the LCT and Doppler electric-field information to the PTD solutions, and again found a potential function to eliminate components of **E** parallel to **B**. The resulting electric-field and Poynting-flux maps are shown in the fifth row of panels in Figure 2, and a scatterplot of the vertical Poynting flux is shown in the lower-right panel of Figure 3.

The linear-correlation coefficient in the Poynting-flux scatterplot is not significantly improved by adding the LCT results to the Doppler results, but the slope of the fit (determined by using IDL's LADFIT function) is somewhat better. Further, examining the maps of E_x and E_y shows a reduction in artifacts in the behavior of the recovered electric-field components, compared to the Doppler and LCT cases. We conclude that at least for this simulation, which exhibited strong flux emergence, most of the additional useful information beyond solutions to Faraday's law is contained within the Doppler velocity measurements, with some additional improvement when non-inductive LCT-derived electric fields are added.

Finally, we wish to add a comment about solutions to the PTD equations themselves. The PTD solutions used in this article did not use FFT solutions for $\dot{\mathcal{B}}$ and $\dot{\mathcal{J}}$, even though the simulations are periodic, but instead used Neumann boundary conditions for $\dot{\mathcal{B}}$ for the reasons described in Section 2.2 of Fisher *et al.* (2010). For the current study, we compared the results of using FFT solutions of the PTD equations with those shown in the figures in this article, and found noticeable degradations in the fits of the model Poynting fluxes to the actual model values. If one is interested in the most accurate reconstruction of the vertical Poynting flux, we recommend not using FFT solutions of the PTD equations.

5. Discussion and Conclusions

We review how the PTD solutions of Faraday's law for **E** can be found using temporal sequences of vector magnetograms, which can be obtained with the HMI instrument on NASA's SDO mission. We discuss why these solutions are under-determined, and we consider the importance of determining the contributions to the electric field that can be derived from a scalar potential.

We demonstrate, using simulation data where the true electric field is known, that knowledge of the vertical-velocity field (obtainable by Doppler measurements) can provide important information about the electric field. When this information is combined with the PTD solutions of Faraday's law, dramatically more accurate recovery of the true electric field is possible. We find that additional information about flows from local correlation-tracking methods can also be combined with the PTD solutions, but the additional information is significantly less important than that from the Doppler measurements. We are able to quantitatively reconstruct the electromagnetic Poynting flux in the simulations by using our combination of the PTD solutions and those from Doppler measurements.

This "proof-of-concept" demonstration argues strongly for the development of electric-field and Poynting-flux tools to be used routinely in the analysis of HMI vector magnetic-field measurements. Routinely available Poynting-flux maps will be useful for scientific studies of flare-energy buildup, understanding the flow of magnetic energy in the solar atmosphere prior to CME initiation, and will aid in understanding the flow of energy that heats the corona. Further, the PTD formalism for the magnetic field itself (Equation (5)) allows for a straight-forward decomposition of the Poynting flux into changes in the potential-field energy, and the flux of free magnetic energy (see Welsch, 2006 and the end of Section 2.1 of Fisher *et al.*, 2010). The flux of free magnetic energy is especially important in determining how energy builds up in flare-productive active regions.

To find solutions for **E** and the Poynting flux **S** using the PTD formalism plus Doppler measurements requires only the solution of three two-dimensional Poisson equations. While real vector-magnetogram patches will not have periodic boundary conditions (as were employed in this article), straightforward numerical techniques exist to solve these equations routinely. Preliminary investigations also indicate that generalizing the PTD solutions and

Doppler measurements to cases of non-normal viewing angle will be straightforward. In our opinion, the major obstacle that remains before such solutions can be routinely applied to the HMI data is a detailed understanding of how measurement errors and disambiguation errors in the vector magnetograms will affect the solutions, and how the effects of these errors are best ameliorated.

Acknowledgements   This research was funded by the NASA Heliophysics Theory Program (grants NNX08AI56G and NNX11AJ65G), the NASA Living-With-a-Star TR&T Program (grant NNX08AQ30G), by the NSF SHINE program (grants ATM0551084 and ATM0752597), and support from NSF's AGS Program (grant ATM0641303) for our participation in the University of Michigan's CCHM Project. The authors are grateful to US taxpayers for providing the funds necessary to perform this work. The authors wish to acknowledge Dick Canfield for his pioneering work in the use of vector magnetograms in solar physics. The inspiration for the work described here can be traced to a Solar MURI workshop held at UC Berkeley in 2002, in which Dick Canfield played a major role in defining long-term research goals for the use of vector magnetograms in quantitative models of the Sun's atmosphere.

References

Chae, J., Sakurai, T.: 2008, A test of three optical flow techniques – LCT, DAVE, and NAVE. *Astrophys. J.* **689**, 593 – 612. doi:10.1086/592761.

Fisher, G.H., Welsch, B.T.: 2008, FLCT: A fast, efficient method for performing local correlation tracking. In: Howe, R., Komm, R.W., Balasubramaniam, K.S., Petrie, G.J.D. (eds.) *Subsurface and Atmospheric Influences on Solar Activity* **CS-383**, Astron. Soc. Pac., San Francisco, 373 – 380 (also arXiv:0712.4289).

Fisher, G.H., Welsch, B.T., Abbett, W.P., Bercik, D.J.: 2010, Estimating electric fields from vector magnetogram sequences. *Astrophys. J.* **715**, 242 – 259. doi:10.1088/0004-637X/715/1/242.

Hurlburt, N.E., Schrijver, C.J., Shine, R.A., Title, A.M.: 1995, Simulated MDI observations of convection. In: Hoeksema, J.T., Domingo, V., Fleck, B., Battrick, B. (eds.) *Helioseismology* **SP-376**, ESA, Noordwijk, 239.

Kusano, K., Maeshiro, T., Yokoyama, T., Sakurai, T.: 2002, Measurement of magnetic helicity injection and free energy loading into the solar corona. *Astrophys. J.* **577**, 501 – 512.

Longcope, D.W.: 2004, Inferring a photospheric velocity field from a sequence of vector magnetograms: the minimum energy fit. *Astrophys. J.* **612**, 1181 – 1192.

November, L.J., Simon, G.W.: 1988, Precise proper-motion measurement of solar granulation. *Astrophys. J.* **333**, 427 – 442.

Potts, H.E., Barrett, R.K., Diver, D.A.: 2004, Balltracking: An highly efficient method for tracking flow fields. *Astron. Astrophys.* **424**, 253 – 262. doi:10.1051/0004-6361:20035891.

Ravindra, B., Longcope, D.W., Abbett, W.P.: 2008, Inferring photospheric velocity fields using a combination of minimum energy fit, local correlation tracking, and Doppler velocity. *Astrophys. J.* **677**, 751 – 768. doi:10.1086/528363.

Scherrer, P.H., The HMI Team: 2005, The helioseismic and magnetic imager for the solar dynamics observatory. In: *AGU Spring Meeting Abstracts*, 43-05.

Schuck, P.W.: 2006, Tracking magnetic footpoints with the magnetic induction equation. *Astrophys. J.* **646**, 1358 – 1391. doi:10.1086/505015.

Schuck, P.W.: 2008, Tracking vector magnetograms with the magnetic induction equation. *Astrophys. J.* **683**, 1134 – 1152. doi:10.1086/589434.

Schuck, P.W., Sun, X., Muglach, K., Hoeksema, J.T.: 2010, Tracking vector magnetograms from the solar dynamics observatory. In: *AGU Fall Meeting Abstracts*, A7.

Title, A.M., Hurlburt, N.E., Schrijver, C.J., Shine, R.A., Tarbell, T.: 1995, Observations of convection. In: Hoeksema, J.T., Domingo, V., Fleck, B., Battrick, B. (eds.) *Helioseismology* **SP-376**, ESA, Noordwijk, 113.

Welsch, B.T.: 2006, Magnetic flux cancellation and coronal magnetic energy. *Astrophys. J.* **638**, 1101 – 1109. doi:10.1086/498638.

Welsch, B.T., Fisher, G.H., Abbett, W.P., Regnier, S.: 2004, ILCT: Recovering photospheric velocities from magnetograms by combining the induction equation with local correlation tracking. *Astrophys. J.* **610**, 1148 – 1156. doi:10.1086/421767.

Welsch, B.T., Abbett, W.P., DeRosa, M.L., Fisher, G.H., Georgoulis, K., Kusano, M.K., Longcope, D.W., Ravindra, B., Schuck, P.W.: 2007, Tests and comparisons of velocity inversion techniques. *Astrophys. J.* **670**, 1434 – 1452.

Solar Phys (2012) 277:165–183
DOI 10.1007/s11207-011-9786-6

SOLAR FLARE MAGNETIC FIELDS AND PLASMAS

Predictions of Energy and Helicity in Four Major Eruptive Solar Flares

**Maria D. Kazachenko · Richard C. Canfield ·
Dana W. Longcope · Jiong Qiu**

Received: 29 November 2010 / Accepted: 9 May 2011 / Published online: 1 July 2011
© Springer Science+Business Media B.V. 2011

Abstract In order to better understand the solar genesis of interplanetary magnetic clouds (MCs), we model the magnetic and topological properties of four large eruptive solar flares and relate them to observations. We use the three-dimensional *Minimum Current Corona* model (Longcope, 1996, *Solar Phys.* **169**, 91) and observations of pre-flare photospheric magnetic field and flare ribbons to derive values of reconnected magnetic flux, flare energy, flux rope helicity, and orientation of the flux-rope poloidal field. We compare model predictions of those quantities to flare and MC observations, and within the estimated uncertainties of the methods used find the following: The predicted model reconnection fluxes are equal to or lower than the reconnection fluxes inferred from the observed ribbon motions. Both observed and model reconnection fluxes match the MC poloidal fluxes. The predicted flux-rope helicities match the MC helicities. The predicted free energies lie between the observed energies and the estimated total flare luminosities. The direction of the leading edge of the MC's poloidal field is aligned with the poloidal field of the flux rope in the AR rather than the global dipole field. These findings compel us to believe that magnetic clouds associated with these four solar flares are formed by low-corona magnetic reconnection during the eruption, rather than eruption of pre-existing structures in the corona or formation in the upper corona with participation of the global magnetic field. We also note that since all four flares occurred in active regions without significant pre-flare flux emergence and cancelation, the energy and helicity that we find are stored by shearing and rotating motions, which are sufficient to account for the observed radiative flare energy and MC helicity.

Keywords Flares, relation to magnetic field · Helicity, magnetic · Flares, models

Solar Flare Magnetic Fields and Plasmas
Guest Editors: Y. Fan and G.H. Fisher

M.D. Kazachenko (✉) · R.C. Canfield · D.W. Longcope · J. Qiu
Montana State University, Bozeman, MT, USA
e-mail: kazachenko@ssl.berkeley.edu

Present address:
M.D. Kazachenko
Space Sciences Laboratory, University of California, Berkeley, CA, USA

1. Introduction

Coronal mass ejections (CMEs) expel plasma, magnetic flux, and helicity from the Sun into the interplanetary medium. At 1 AU, in the interplanetary medium, CMEs appear as interplanetary coronal mass ejections (ICMEs). At least one third (Gosling, 1990) or perhaps a larger fraction (Webb *et al.*, 2000) of the ICMEs observed *in situ* are magnetic clouds (MCs) (Burlaga *et al.*, 1981), coherent "flux-rope" structures characterized by low proton temperature and strong magnetic field whose direction typically rotates smoothly as they pass the spacecraft.

MCs originate from eruptions of both quiescent filaments and active regions (ARs). The 3D magnetic models and geomagnetic relationships are better understood for filament eruptions than for ARs (Marubashi, 1986; Bothmer and Schwenn, 1998; Zhao and Hoeksema, 1998; Yurchyshyn *et al.*, 2001). However, the most geoeffective MCs originate from ARs (Gopalswamy *et al.*, 2010). In this article we focus exclusively on the latter.

Comparison of the properties of MCs with those of their related ARs clarifies our understanding of both domains. Assuming MCs to be twisted flux ropes in magnetic equilibrium, several authors have succeeded in inferring global properties such as MC axis orientation, net magnetic flux, and magnetic helicity (see review by Démoulin, 2008). For a sample of twelve MCs, Leamon *et al.* (2004) found that the percentage of MC poloidal flux relative to unsigned vertical AR flux varied widely: from 1% to 300%. For one MC, Luoni *et al.* (2005) did a similar study and found a factor of ten times lower flux in the MC than the AR, in agreement with other previous studies (Démoulin *et al.*, 2002; Green *et al.*, 2002). More recently, for a sample of nine MCs, Qiu *et al.* (2007) found that the MC poloidal flux matches the "observed" reconnection flux, *i.e.* reconnection flux in the two-ribbon flare associated with it, and the toroidal flux is a fraction of the reconnection flux. The Qiu *et al.* (2007) results may be interpreted as evidence of formation of the helical structure of magnetic-flux ropes by reconnection, in the course of which magnetic flux, as well as helicity, is transported into the flux rope.

The other quantity that is very useful for relating MCs to their associated flares is magnetic helicity, which describes how sheared and twisted the magnetic field is compared to its lowest energy state (Berger, 1999; Démoulin and Pariat, 2007). Since helicity is approximately conserved in the solar atmosphere and the heliosphere (Berger and Field, 1984), it is a very powerful quantity for linking solar and interplanetary phenomena. For six ARs, Nindos, Zhang, and Zhang (2003) found that photospheric helicity injection in the whole AR is comparable with the MC helicity. However, it is worth remembering that this approach is simplified, since the liftoff of the flux rope does not remove all of the helicity available in the AR (Mackay and van Ballegooijen, 2006; Gibson and Fan, 2008). Mandrini *et al.* (2005) and Luoni *et al.* (2005) compared, respectively, the helicity released from a very small AR and a very large AR, with the helicity of their associated MC. They found a very good agreement in the values (small AR with small MC, and large AR with large MC), despite a difference of three orders of magnitude between the smaller and the larger events.

There exist two basic ideas about the solar origin of magnetic clouds: MCs are formed either globally or locally. In the *global* picture, the MC topology is defined by the overall dipolar magnetic field of the Sun (Crooker, 2000). In this case, the field lines of the helmet-streamer belt become the outermost coils of the MC through reconnection behind the CME as it lifts off. Hence the leading field direction of the magnetic cloud tends to follow that of the large-scale solar dipole, reversing at solar maximum (Mulligan, Russell, and Luhmann, 1998; Bothmer and Schwenn, 1998; Li *et al.*, 2011). In the *local* picture, on the other hand, the flux rope is formed within the AR and its properties are defined by properties of the AR. We can categorize the "local" models into two sub-classes: In the first, the magnetic-flux rope emerges from beneath the photosphere into the corona (Chen, 1989; Low, 1994; Leka *et al.*, 1996; Abbett and Fisher, 2003; Fan and Gibson, 2004). In this scenario the flux ropes formed may maintain stability for a relatively long time prior to the explosive loss of equilibrium (Forbes and Priest, 1995; Lin, Raymond, and van Ballegooijen, 2004) or a breakout type reconnection that opens up the overlying flux rope of opposite polarities (Antiochos, Devore, and Klimchuk, 1999). Such a flux rope is therefore *pre-existing* before its expulsion into interplanetary space. In the second case the flux rope is formed *in situ* by magnetic reconnection. The magnetic reconnection suddenly reorganizes the field configuration in favor of expulsion of the "*in-situ*" formed magnetic-flux rope out of the solar atmosphere. The results of Qiu *et al.* (2007) support this case. Qiu *et al.* (2007) found that the reconnection flux from observations of flare-ribbon evolution is greater than toroidal flux of the MC but comparable and proportional to its poloidal flux, regardless of the presence of filament eruption. Their conclusion agrees with the inference from the study by Leamon *et al.* (2004), although through a very different approach.

Our working hypothesis is that MCs associated with the ARs originate from the ejection of locally *in-situ* formed flux ropes. In this case shearing and rotation of the photosphere magnetic-flux concentration before the flare lead to the build-up of magnetic stress, which is removed during the flare by reconnection. As a result a magnetic-flux rope is formed and erupts, producing a MC.

To test our hypothesis we apply a quantitative, non-potential, self-consistent model, the *Minimum Current Corona* (MCC) model (Longcope, 1996, 2001), to predict the properties of the *in-situ* formed flux rope in four two-ribbon flares. Using the MCC model with SOHO/MDI magnetogram sequences we construct a three-dimensional model of the pre-flare magnetic-field topology and make quantitative predictions of the amount of magnetic flux that reconnects in the flare, the magnetic self-helicity of the flux rope created, and the minimum energy release the topological change would yield. We then compare the predicted flare helicity and energy to MC helicity and flare energy, inferred from fitting the magnetic cloud (*Wind*, ACE) and GOES observations correspondingly. We compare the predicted reconnected flux to the amount of photospheric flux swept up by the flare ribbons using TRACE 1600 Å data and the poloidal MC flux inferred from fitting the magnetic-cloud observations. We find that for the four studied flares our results support, from the point of view of flux, energy and helicity, the scenario in which the MC progenitor is a helical flux rope formed *in situ* by magnetic reconnection in the low corona immediately before its expulsion into interplanetary space. We also find that MC topology is defined by the local AR structure rather than the overall dipolar magnetic field of the Sun in the events studied.

This article is organized as follows: in Section 2 we describe the methods and uncertainties of our analysis. In Section 3 we describe the four flares studied, the flux and helicity of the ARs in which the flares occurred, and the magnetogram sequence during the build-up time. In Section 4 we discuss our results, and in Section 5 summarize our conclusions.

2. Methods: Calculating Reconnection Flux, Energy and Helicity

In this section we describe the methods that we use to predict the reconnection flux, energy, and helicity from SOHO/MDI magnetogram sequences and the *Minimum Current Corona* model (see Section 2.1) and determine observed values of these quantities from GOES, TRACE, ACE, and *Wind* observations (see Section 2.2)

Table 1 Flare and AR Properties (see Section 3). Number is the NOAA number of the AR associated with the flare; Φ_{AR}, in units of 10^{22} Mx, is the AR's unsigned magnetic flux; and H_{AR}, in units of 10^{42} Mx^2, is helicity injected into the AR during the magnetogram sequence starting at t_0 and ending at t_{flare}.

i	Flare			Active region			M-gram sequence	
	Date	Time	Class	Number	Φ_{AR}	H_{AR}	t_0	t_{flare}
1	13 May 2005	16:57	M8	10759	2.0	-12 ± 1.2	11 May 23:59	13 May 16:03
2	7 Nov 2004	16:06	X2	10696	2.1	-15 ± 1.5	6 Nov 00:03	7 Nov 16:03
3	14 Jul 2000	10:03	X6	09077	3.4	-27 ± 2.7	12 Jul 14:27	14 Jul 09:36
4	28 Oct 2003	11:10	X17	10486	7.5	-140 ± 14.0	26 Oct 12:00	28 Oct 10:00

2.1. Minimum Current Corona Model

The key improvement of our study relative to Leamon *et al.* (2004) and Qiu *et al.* (2007) is the use of the *Minimum Current Corona* model, which allows us to estimate the energy and helicity associated with the *in-situ* formed flux rope (Longcope, 1996, 2001). The MCC model extends the basic elements of the CSHKP (Carmichael, 1964; Sturrock, 1968; Hirayama, 1974; Kopp and Pneuman, 1976) two-ribbon flare scenario to three dimensions, including the shearing of an AR along its polarity-inversion line (PIL) to build up stress. After this pre-flare stress build-up, the MCC model quantifies the result of eliminating some or all of the stress and creating a twisted flux rope overlying the AR, through magnetic reconnection.

To describe the evolution of the pre-flare photospheric motions that lead to stress build-up we use a sequence of SOHO/MDI full-disk magnetograms (Scherrer *et al.*, 1995). As the starting point we take t_0, right after the end of a large flare, which we call the zero-flare. We assume that at t_0 the AR's magnetic field becomes fully relaxed. As the ending time we take t_{flare}, right before the time when the flare of study occurred but avoiding artifacts associated with the onset of the flare brightening (Qiu and Gary, 2003). To achieve the maximum energy release, the field reconnecting during the flare would need to relax to its potential state, hence we assume the field to be potential at t_{flare}. As a result, we form a sequence of magnetograms, which covers Δt hours of stress build-up prior to the flare (see Table 1).

For quantitative analysis of the pre-flare magnetic field we divide each magnetogram into a set of unipolar *partitions* and then into unipolar magnetic charges (*e.g.* see partitioned magnetogram in Appendix, Figure 4). Firstly, for all successive pairs of magnetograms we derive a local correlation tracking (LCT) velocity field (November and Simon, 1988; Chae, 2001) and then group pixels into individual partitions that have persistent labels. In the second step we represent each magnetic partition with a magnetic *point charge* (or magnetic point source) which has the flux of the partition and is located at its center of flux. Finally, using the LCT velocity field we calculate the helicity injected by the motions of the magnetic point charges of the whole AR (H_{AR}, see Table 1 and Longcope *et al.* (2007)). We make sure that the amount of helicity injected by the motions of the continuous photospheric partitions matches the helicity injected by the motions of the magnetic point charges. Their equality gives us confidence that the centroid motions of the point charges accurately capture helicity injection. As has been shown by Chae (2007), computing the vector potential via the Fourier approach of Chae (2001), as we choose to do, reults in the higher values (10%) in the helicity flux compared to that from the approach by Pariat, Démoulin, and Berger (2005). In addition, the LCT method that we use yields systematically lower values than the DAVE velocity-inversion algorithm with a difference in helicity flux of less than $\approx$ 10% (Welsch

Table 2 Flare physical properties: MCC model predictions *vs.* observations: predicted $\Phi_{r,MCC}$ and inferred from observations $\Phi_{r,ribbon}$ reconnection fluxes and MC poloidal fluxes $\Phi_{p,MC}$, predicted $\mathcal{E}_{MCC}$ and observed $\mathcal{E}_{GOES}$ energy values, predicted H_{MCC} and observed H_{MC} helicity values (see Section 3).

i	$\Phi_{r,MCC}$ 10^{21} Mx	$\Phi_{r,ribbon}$ 10^{21} Mx	$\Phi_{p,MC}$ 10^{21} Mx	$\mathcal{E}_{MCC}$ 10^{31} ergs	$\mathcal{E}_{GOES}$ 10^{31} ergs	H_{MCC} 10^{42} Mx^2	H_{MC} 10^{42} Mx^2
1	2.8 ± 0.4	4.1 ± 0.4	6.3 ± 4.2	1.0 ± 0.3	3.1 ± 0.6	-7.0 ± 1.2	-7.5 ± 5.0
2	5.4 ± 0.8	4.8 ± 0.5	5.25 ± 3.5	6.4 ± 1.8	2.0 ± 0.1	-5.0 ± 0.6	-8.3 ± 5.5
3	6.0 ± 0.9	12.8 ± 3.0	9.9 ± 6.6	9.1 ± 2.6	10.1 ± 2.1	-20.1 ± 3.6	-22.5 ± 15.0
4	15.0 ± 2.6	23.0 ± 7.0	18.0 ± 12.0	18.0 ± 5.2	13.6 ± 0.6	-48.0 ± 8.6	-45.0 ± 30.0

et al., 2007). Those two effects result in an uncertainty of 10% in the H_{AR} value which we take into account.

The MCC model characterizes the changes in the pre-flare magnetic field purely in terms of the changes in the *magnetic domains*: volumes of field lines connecting pairs of opposite point charges. Replacing each partition with a single magnetic point charge as we chose to do results in values of domain fluxes that are only slightly different from the actual domain fluxes (Longcope, Barnes, and Beveridge, 2009). As magnetic charges move, the magnetic field, first relaxed by the zero-flare, becomes increasingly stressed and hence non-potential. Under the assumption that no reconnection, flux emergence, or flux cancelation occur between the zero-flare and the flare of interest, the domain fluxes could not have changed. (Note that for our analysis we selected only flares associated with ARs with no significant flux emergence or cancelation during the period between t_0 and t_{flare}.) To provide both the domain flux conservation and the increasing field non-potentiality, the MCC model includes currents only on the intersections between the domain boundaries, called *separators*. In this way the lack of reconnection leads to storage of free magnetic energy, energy above that of the potential field, which could then be released by reconnection in the flare. To achieve the maximum energy release, the field inside the domains associated with the flare (*flaring domains*) would need to relax to its potential state. Thus to find the reconnection flux we first need to find the flaring domains and then calculate the changes in their domain flux from t_0 to t_{flare}. More specifically, we first overlay the magnetic point charges rotated to the time of the TRACE 1600 Å flare ribbons onto the ribbons image (*e.g.* of Figure 5 in Appendix) and then use a Monte-Carlo method (Barnes, Longcope, and Leka, 2005) to find the fluxes of the flaring domains at t_0 and t_{flare}. Finally, we separately sum up the absolute values of all the positive and negative changes in the domain fluxes to calculate the model reconnection flux [$\Phi_{r,MCC}$, see Table 2]. This is the model estimate of the net flux transfer that must occur in the two-ribbon flare through the flare reconnection.

To find the *flaring separators*, we find the topology of the magnetic field at t_{flare} and select those separators that connect nulls that are located on the flare ribbons. Through the MCC model, the changes in the domain fluxes under those flaring separators allow us to calculate current, free energy, and helicity liberated on each separator (for a detailed description of the method see Longcope (1996) and Appendix B of Kazachenko *et al.* (2009)). The total model energy [$\mathcal{E}_{MCC}$] released during the flare is a sum of energies released at each flaring separator. It is a lower bound on the energy stored by the pre-flare motions, since the MCC model uses the point-charge representation and hence applies a smaller number of constraints than point-for-point line tying. It can be shown that the energy stored by ideal, line-tied, quasi-static evolution will always exceed the energy of the corresponding flux-constrained equilibria (Longcope and Magara, 2004). The total mutual helicity injected on all flaring separators is a sum of the helicities injected on each flaring separator. However, the liftoff of the flux rope does not remove all of the helicity available in the flux rope (Mackay and van Ballegooijen, 2006). For an MHD-simulated eruption, Gibson and Fan (2008) found that 41% of the helicity is lost with the escaping rope, while 59% remains. For simplicity, we assume that 50% of the total mutual helicity from the MCC model ends up as self-helicity of the flux rope created by reconnection: $H_{MCC} = \sum H_i/2$. Finally we note that the MCC model depends on the way that we partition the magnetogram sequence (Beveridge and Longcope, 2006). We experimented with different values of saddle points in the partitioning and apodizing windows in the LCT and found that this contributes an uncertainty in the MCC reconnection flux, MCC energy, and MCC helicity that we include in Table 2.

2.2. Flare and MC Observations

For comparison with the predictions of the MCC model, we must infer values of reconnection flux $[\Phi_{r,ribbon}]$ from the observations, MC poloidal flux $[\Phi_{p,MC}]$, energy $[\mathcal{E}_{GOES}]$, and helicity $[H_{MC}]$ (see Table 2).

To infer values of reconnection flux $[\Phi_{r,ribbon}]$ from the observations, we use flare-ribbon motion (Poletto and Kopp, 1986; Fletcher and Hudson, 2001) observed in 1600 Å images from TRACE. To find the total magnetic flux swept out by a moving ribbon, we count all pixels that brightened during any period of the flare and then integrate the unsigned magnetic flux encompassed by the entire area taking into account the height of the ribbon's formation, a $\approx 20\%$ correction (Qiu *et al.*, 2007). The uncertainties in the $\Phi_{r,ribbon}$ are estimated by artificial misalignment between the MDI and TRACE data, ribbon-edge uncertainty, and inclusion of transient non-ribbon features with the ribbon areas. To quantify the misalignment contribution, we perform a set of trials whereby magnetogram and 1600 Å images are offset by up to two MDI pixels. To find the uncertainty due to ribbon-edge identification, we perform the calculation for different ribbon-edge cutoff values ranging from six to ten times the background intensity. We also compare the MCC reconnection flux $[\Phi_{r,MCC}]$ to the poloidal MC flux $[\Phi_{p,MC}]$ derived from fits to the *in situ* MC ACE/*Wind* observations using the Grad–Shafranov reconstruction method (Hu and Sonnerup, 2001). As Qiu *et al.* (2007) showed from observations, $\Phi_{r,ribbon} \approx \Phi_{p,MC}$. Hence if the MCC model captures the reconnection flux correctly, $\Phi_{r,MCC}$ should match $\Phi_{p,MC}$ unless reconnection of the ICME with the ambient solar wind makes an important contribution (Dasso *et al.*, 2006).

During the flare, the magnetic free energy that has been slowly stored by photospheric motions $[\mathcal{E}_{MCC}]$ is rapidly released by reconnection and then dissipated. We estimate energy losses not only due to radiation $[\mathcal{E}_r]$, as Kazachenko *et al.* (2009) and Kazachenko *et al.* (2010) did, but also due to conductive cooling $[\mathcal{E}_c]$ and the enthalpy flux $[\mathcal{E}_{ent}]$, which in some numerical cases is as large as radiative energy losses (Bradshaw and Cargill, 2010). Since it is not clear whether the source for the CME kinetic energy is the magnetic free energy stored in the active region and not the energy stored, *e.g.*, in the interplanetary current sheet, we neglect the energy carried away by the CME.

To quantify the three components of $\mathcal{E}_{GOES} = \mathcal{E}_r + \mathcal{E}_c + \mathcal{E}_{ent}$ (see Table 3) we use GOES analysis software in SolarSoft and the observed GOES X-ray fluxes in the two channels (1–8 Å and 0.5–4 Å). Those provide an estimate of the plasma temperature T and emission measure $EM = n_e^2 V$, where n_e is the electron density and V is the emitting volume. Radiative energy losses $\mathcal{E}_r$ depend on the emission measure, temperature and composition of emitting plasma. We find their magnitude using the temperature-dependent Mewe radiative-loss function (Mewe, Gronenschild, and van den Oord, 1985). To calculate the conductive energy losses $[\mathcal{E}_c]$ we integrate the conductive energy loss rate to

Table 3 Observed energy budget, in units of 10^{31} ergs: radiative losses $\mathcal{E}_r$, conductive losses $\mathcal{E}_c$, enthalpy fluxes $\mathcal{E}_{ent}$, total energy $\mathcal{E}_{GOES} = \mathcal{E}_c + \mathcal{E}_r + \mathcal{E}_{ent}$ and estimated value for flare luminosity $\mathcal{E}_{FL} \approx (3.15 \pm 1.05) \times \mathcal{E}_{GOES}$. For comparison with the observations, the predicted model energy $\mathcal{E}_{MCC}$ is given (see Section 2.2 and Section 3).

i	L (Mm)	$\mathcal{E}_r$	$\mathcal{E}_c$	$\mathcal{E}_{ent}$	$\mathcal{E}_{GOES}$	$\mathcal{E}_{FL}$	$\mathcal{E}_{MCC}$
1	145 ± 31	1.0	0.45 ± 0.05	1.6 ± 0.6	3.1 ± 0.6	10.3 ± 5.1	1.0 ± 0.3
2	43 ± 20	0.9	0.4 ± 0.1	0.7 ± 0.1	2.0 ± 0.1	6.4 ± 2.4	6.4 ± 1.8
3	151 ± 50	2.5	2.6 ± 0.2	4.9 ± 1.9	10.1 ± 2.1	34.0 ± 17.2	9.1 ± 2.6
4	107 ± 18	5.3	2.75 ± 1.35	5.5 ± 0.7	13.6 ± 0.6	43.5 ± 16.1	18.0 ± 5.2

the chromosphere [$P_{cond} = U_{th}/\tau_{cond}$] where U_{th} is thermal energy content of the plasma, $U_{th} = 3n_e kTV = 3kT\sqrt{EM \times V}$ and τ_{cond} is the cooling time scale

$$\tau_{cond} \approx \frac{3kn_e(L/2)^2}{\kappa_0 T^{5/2}}, \tag{1}$$

for a loop of full length L, Boltzmann constant k, and Spitzer conductivity $\kappa_0 \approx 10^{-6}$ (Longcope *et al.*, 2010). We quantify the volume of the emitting material [V] by assuming that $\mathcal{E}_c \approx \mathcal{E}_r$ at late times, as should be the case in a static equilibrium (Rosner, Tucker, and Vaiana, 1978; Vesecky, Antiochos, and Underwood, 1979). From the volume [V] and emission measure [EM] we derive the electron density $n_e = \sqrt{EM/V}$. For the loop length [L] we use the distribution of the lengths of the flaring separators (with the energy weights) whose geometrical properties are found from the coronal magnetic topology at t_{flare} (Section 2.1). The mean and standard deviations of the lengths of the flaring separators yield the mean and the standard deviation of the values of $\mathcal{E}_c$. Finally, we estimate the enthalpy flux [$\mathcal{E}_{ent}$] using model calculations by Bradshaw and Cargill (2010). From Tables 1 and 2 in Bradshaw and Cargill (2010) and the loop lengths of the flaring separators, we first derive a coefficient [δ] which describes the ratio between the radiative cooling and the enthalpy-flux time scales and then the enthalpy flux itself.

We get an additional idea for the value of the uncertainty in $\mathcal{E}_{GOES}$ by comparing it to the flare luminosity [FL, $\mathcal{E}_{FL}$] from the *Total Irradiance Monitor* (TIM) on the *Solar Radiation and Climate Experiment* (SORCE). Unfortunately, FL measurements are not available for the four flares studied here. However, FLs have been measured for four other large ($> X10$) solar flares (see Table 2 in Woods, Kopp, and Chamberlin (2006)), for which we may calculate $\mathcal{E}_{GOES}$. For these four flares we find that the FLs are approximately two to four times larger than the $\mathcal{E}_{GOES}$. We use this scaling to limit our energy estimates from above (see Table 3): $\mathcal{E}_{FL} \approx (3.15 \pm 1.05) \times \mathcal{E}_{GOES}$.

Finally, we compare the model MCC flux-rope helicity with the helicity of the magnetic cloud associated with the flare [H_{MC}]. We calculate H_{MC} applying the Grad–Shafranov method (Hu and Sonnerup, 2001) to the ACE/*Wind* MC observations. There are several uncertainties and limitations in the determination of H_{MC}, which are as well applicable to MC poloidal-flux calculations. First, the inferred value of H_{MC} is model-dependent: *e.g.* within the cylindrical hypothesis, force-free and non-force-free models give helicities values that differ by up to 30% (Dasso *et al.*, 2003, 2006). However, this variation remains small compared to the variation of helicity values computed for different MCs (Gulisano *et al.*, 2005). Second, the MC boundaries can be defined by several criteria, which do not always agree. This introduces an uncertainty on the magnetic helicity which can be comparable to the uncertainty obtained with different models (Dasso *et al.*, 2006). Finally, the distribution of the twist along the flux rope as well as the length of the flux rope are generally not known. So far only in one case has the length of the flux rope $L_{MC} = 2.5$ AU been determined precisely from impulsive electron events and solar type III radio bursts (Larson *et al.*, 1997). We take the value of 0.5 AU as the lower limit of L_{MC} (DeVore, 2000) and 2.5 AU as the upper limit of L_{MC} (Larson *et al.*, 1997). Such choices of the lower and upper limits of L_{MC} would change poloidal MC flux and MC helicity to vary between roughly half and twice the measured values.

3. Data: Flares Studied

We apply the methods described in the previous section to four large eruptive flares (Table 1). This number of events is limited by several necessary flare selection criteria. Firstly, we selected only events that have good observations of both the flare and the MC. Secondly, except for the 13 May 2005 flare, we selected only ARs where two successive flares larger than M-class were present, in order to make plausible our assumption of initial relaxation of the AR's magnetic field to potential state. Thirdly, both the flare of study and the zero-flare should happen no farther than 40° from the central meridian so that the stress build-up could be observed. Finally, we selected only flares associated with ARs with no significant flux emergence or cancelation during the period between t_0 and t_{flare}.

Our topological analysis using the MCC model has been applied previously for three of the four flares: M8 flare on 13 May 2005 (Kazachenko 2009), X2 flare on 7 November 2004 (Longcope *et al.*, 2007), X17 flare on 28 October 2003 (Kazachenko *et al.*, 2010). The results of the MCC analysis for the X5.7 flare on 14 July 2000 are described in this article for the first time (see Appendix). In Table 1 we list the flare number i in this work, date, time and X-ray class of each flare; the NOAA number of the AR associated with the flare, AR's unsigned magnetic flux [Φ_{AR}] and helicity injected into the AR during the magnetogram sequence [H_{AR}]; start [t_0] and end time [t_{flare}] of the magnetogram sequence. The flares are sorted by X-ray class. In Table 2 we compare MCC-model predicted physical properties with the observations: predicted and inferred from the ribbon-motion reconnection fluxes and MC poloidal fluxes, predicted and observed from the GOES observations energy values, predicted and observed from the *Wind*/ACE observations helicities. Finally in Table 3 we detail the observed energy budget for each flare.

The first flare listed in Table 1 is the M8 flare that occurred on 13 May 2005 in NOAA 10759 (Kazachenko *et al.*, 2009; Yurchyshyn *et al.*, 2006; Liu *et al.*, 2007; Jing *et al.*, 2007; Liu, Zhang, and Zhang, 2008). NOAA 10759 had a large positive sunspot which contained more than a half of the total positive flux of the AR and rotated with the rate of $0.85 \pm 0.13°$ per hour during the 40 hours before the flare (Kazachenko *et al.*, 2009). Such fast rotation, along with the fact that the spin-helicity flux is proportional to the magnetic flux squared, makes the effect of sunspot rotation dominant in the helicity budget of the whole AR. As for the flare itself, the rotation of the sunspot produced three times more energy and magnetic helicity than in the hypothetical case in which the sunspot does not rotate; the inclusion of sunspot rotation in the analysis brings the model into substantially better agreement with GOES and interplanetary magnetic-cloud observations. Rotation is energetically important in the flare and alone can store sufficient energy to power this M8 flare.

The second flare in Table 1 is the X2 flare on 7 November 2004 (Longcope *et al.*, 2007). The start time was plausibly taken to be that of an M9.3 flare which occurred 40 hours before the flare of interest. The MCC model predicts a value of the flux needed to be reconnected in the flare that compares favorably with the flux swept up by the flare ribbons. The MCC model places a lower bound on the energy stored by the 40-hour build-up shearing motions that is at least three times larger than the observed energy losses. The helicity assigned to the flux rope that is assumed in the model is comparable to the magnetic cloud helicity. Note that our estimate for H_{MCC} in Table 2 is higher than the one in Longcope *et al.* (2007) (see Table 2 in Longcope *et al.* (2007)): we estimate H_{MCC} as a sum of the helicities over all eight flaring separators, while Longcope *et al.* (2007) took a sum over only the three most energetic separators.

The third flare in Table 1 is the X6 flare on 14 July 2000 (Lepping *et al.*, 2001; Yurchyshyn *et al.*, 2001; Fletcher and Hudson, 2001; Masuda, Kosugi, and Hudson, 2001). Our analysis of this flare is described in this article for the first time (see Appendix). We take as the zero-flare an X1.9 flare around 48 hours before the flare of interest. We use the MCC model to find that the released energy is comparable to the observed energy losses. The amount of flux reconnected during the flare according to the model is at least one and a half times smaller than the reconnection flux observed with TRACE. The model estimate for the helicity is comparable with the helicity from the MC observations. No sunspot rotation is associated with the pre-flare evolution.

The fourth event in Table 1, the X17 Halloween flare (Yurchyshyn, Hu, and Abramenko, 2005; Régnier and Priest, 2007; Schrijver *et al.*, 2006; Mandrini *et al.*, 2006; Lynch *et al.*, 2005; Zhang, Liu, and Zhang, 2008), occurred in an AR with a fast-rotating sunspot. We find that the MCC reconnection flux is consistent with the reconnection flux inferred from the observations. We find that the sunspot rotation increases the total AR helicity by $\approx 50\%$. However, in contrast to the flare on 13 May 2005, where rotation is dominant in the energetics, rotation increases the free energy and flux-rope helicity of this flare by only $\approx 10\%$. Shearing motions alone store sufficient energy and helicity to account for the flare energetics and ICME helicity content within their observational uncertainties. Thus this flare demonstrates that the relative importance of shearing and rotation in this flare depends critically on their location within the parent AR topology.

4. Results: MCC Model Predictions *Versus* Observations

The main global property that describes the flare's reconnection is the amount of magnetic flux that participates. Figure 1 shows predicted $[(\Phi_{r,MCC}]$ and observed $[\Phi_{r,ribbon}]$ reconnection fluxes for each event, the AR-average unsigned magnetic flux $[\Phi_{AR}]$ and the poloidal MC flux $[\Phi_{p,MC}]$. The first thing that we notice is that the fraction of the AR magnetic flux that is observed to reconnect during the four flares ranges from 18% to 49%. Secondly, in the second and fourth flares the predicted reconnection flux matches the reconnection flux inferred from the observations, while in the first and third flares the highest probable value of the MCC reconnection flux is lower than the lowest probable value of the observed reconnection flux by 13% and 29% correspondingly. The lower model reconnection flux is likely due to additional reconnections not accounted for in the model. That means that the MCC model captured a lower limit of the amount of magnetic flux that has reconnected in these flares and hence the lower limit on the amount of energy released. Finally, in all four cases the value of poloidal MC flux matches both the observed and model reconnection fluxes, although the uncertainties in $\Phi_{p,MC}$ due to the unknown MC length are quite large. According to the CSHKP model, on which the MCC model builds, reconnection contributes solely to the incremental poloidal component of the flux-rope flux. Therefore, the derived agreement between the poloidal MC flux and the reconnection fluxes means that the flux rope is formed *in situ*.

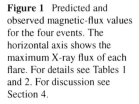

Figure 1 Predicted and observed magnetic-flux values for the four events. The horizontal axis shows the maximum X-ray flux of each flare. For details see Tables 1 and 2. For discussion see Section 4.

Figure 2 Predicted and observed energy values for the four events. For details see Table 3. For discussion see Section 4.

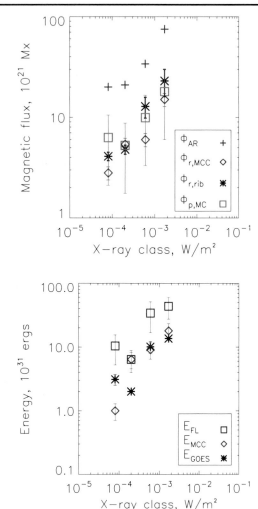

The MCC model gives a lower limit of the free magnetic energy released in each flare (Longcope, 2001). In Figure 2 we compare the predicted MCC model free energy [$\mathcal{E}_{MCC}$, diamonds] with the observed time-integrated sum of radiative and conductive energy losses and the enthalpy flux [$\mathcal{E}_{GOES}$, stars]; we also show the estimated flare luminosity [$\mathcal{E}_{FL}$, blue squares]. Figure 2 indicates that for the third and fourth flares the predicted energy [$\mathcal{E}_{MCC}$] matches the observed energy [$\mathcal{E}_{GOES}$], while for the second flare the predicted energy is around three times larger than $\mathcal{E}_{GOES}$. In all four cases the flare luminosity is higher than both $\mathcal{E}_{GOES}$ and $\mathcal{E}_{MCC}$. This is not surprising, since the MCC model uses a point-charge representation rather than line tying and yields a lower limit on the free energy released in the flare. Summarizing, within the uncertainties, for three flares $i = 2, 3, 4$ the predicted free energy lies between the observed estimate of released energy and the estimated flare luminosity. Only for the 13 May 2005 flare is the model energy lower than the observed estimate. Note, that this flare is the only flare that did not have a zero-flare at t_0. Since rotation is the dominant source of helicity injection in this flare and the rotation rate was around zero before t_0, we believe that our analysis plausibly captures the major source of

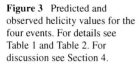

Figure 3 Predicted and observed helicity values for the four events. For details see Table 1 and Table 2. For discussion see Section 4.

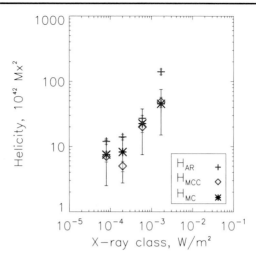

helicity injection in this flare. Nevertheless, the absence of a zero-flare indicates that there might have been additional energy storage before t_0. Hence, our estimate is a lower limit to the reconnection flux and the magnetic energy of this M8 flare.

One must keep in mind that one of the basic assumptions of the MCC model is the potentiality of the magnetic field after the zero-flare. Su, Golub, and Van Ballegooijen (2007) analyzed TRACE observations of 50 X- and M-class, two-ribbon flares and found that 86% of these flares show a general decrease in the shear angle between the main polarity-inversion line and pairs of conjugate bright ribbon kernels. They interpreted this as a relaxation of the field toward a more potential state because of the eruption that carries helicity/current with it, but one can readily argue that a similar decrease in shear angle would be seen if sequentially higher, less-sheared post-flare loops light up with time as the loops cool after reconnection. These results are consequently ambiguous: they may show a decrease in shear, or they may reflect that flares generally do not release all available energy and part of the flux-rope configuration remains. In other words, the MCC model potentiality assumption may mean that additional energy and reconnection flux are stored before the zero-flare.

Because magnetic helicity is approximately conserved in the corona, even in the presence of reconnection, it is instructive to compare H_{MCC} to H_{MC}. Figure 3 shows the relationship between the predicted [H_{MCC}, diamonds] and observed [H_{MC}, stars] values of helicity. The blue + signs show the amount of the helicity for the whole AR [H_{AR}]. In all four cases we get MCC model helicities that are of comparable magnitude and same sign as the observed MC values and are smaller than the helicity of the whole AR. Our analysis shows that pre-flare motions contribute enough stress to account for observed helicity values, however, more accurate estimate of the MC length is required in order to lower the error bars in the MC helicities and improve our understanding of MC–flare relationship. It is interesting that the most energetic of the four flares ($i = 4$, which happened in the southern hemisphere) had the helicity sign opposite to the hemispheric helicity preference (Pevtsov and Balasubramaniam, 2003). Hence its sign cannot be predicted from the global solar properties, but only from a case study like this.

The properties of the magnetic field in a MC are determined by the initial conditions of the eruption, which we derive with the MCC model as well as by how the MC interacts with the interplanetary medium during its travel toward the Earth. Above we found a consistency between the MC flux and predicted model and observed flare-reconnection fluxes, the MC and predicted flare flux-rope helicities, observed and predicted flare energy releases. The agreement between those supports our local *in-situ* formed flux-rope hypothesis.

One more quantity that is frequently compared between the MC and AR flux rope is the direction of the poloidal field. Li *et al.* (2011) found that the poloidal field of MCs with low axis inclination relative to the ecliptic ($\approx$ 40% of all MCs) has a solar-cycle dependence. They note that during the solar minima, the orientation of the leading edge of the MC is predictable: it is the same as the solar dipole field. However, during the maximum and the declining phases, when most of the geoeffective MCs happen, both (north and south) orientations are present, although the global dipole-field orientation of the beginning of the cycle dominates.

It is instructive to consider how the four flares of our study relate to these results. Assuming that the poloidal field in the ejected flux rope is oriented along the flaring separators, we define its orientation relative to the ecliptic plane using the North–South classification: North ($30 < \theta < 90°$, $B_z > 0$, in the solar ecliptic coordinate system) or South ($-90 < \theta < -30°$, $B_z < 0$). We then determine the orientation of the leading edge of the MC poloidal field using the same North–South classification and compare two quantities: the orientation of the poloidal field in the active region and the orientation of the leading edge of the MC poloidal field. We find that the MC produced by the Bastille day flare during the solar maximum has a South-oriented leading MC poloidal field, the same as both the remnant weak dipole orientation and poloidal-field orientation of the flux rope at the Sun. In contrast, the flares that occurred during the declining phase, on 13 May 2005 and 7 November 2004, produced magnetic clouds with South-oriented leading MC poloidal fields, opposite to the direction of the global dipole field, but the same as the poloidal-field orientation predicted for a flux rope in the modeled AR. Finally the MC produced by the Halloween flare lay perpendicular to the ecliptic plane and thus was not relevant to the observed Li *et al.* (2011) rule; however, a good agreement was also found between the directions of the poloidal field in the MC and in the source AR (Yurchyshyn, Hu, and Abramenko, 2005).

Summarizing the above: although there is a tendency for ARs to follow the dipole-field orientation during the solar minimum (Li *et al.*, 2011), during the solar maximum and the declining phase, when the largest MCs occur, the local AR field is important. We find that for the four studied large events, the direction of the leading MC poloidal field is consistent with the poloidal-field orientation in the AR rather than to the global dipole field, in agreement with Leamon, Canfield, and Pevtsov (2002). This implies that the magnetic clouds associated with large ARs inherit the properties of the AR rather than those of the global dipole field, as a result of reconnection in the active region rather than with the surrounding dipole field. Although here we compare the poloidal post-flare arcade field with the poloidal MC field, this supports the conclusion by Yurchyshyn *et al.* (2007), who found that 64% of CMEs are oriented within 45° to the MC axes (MC toroidal field) and 70% of CMEs are oriented within 10° to the toroidal field of EUV post-flare arcades (Yurchyshyn, Abramenko, and Tripathi, 2009). In other words, despite the fact that CME flux ropes may interact significantly with the ambient solar wind (Dasso *et al.*, 2006) or other flux ropes (Gopalswamy *et al.*, 2001), a significant group of MCs reflects the magnetic field orientation of the source regions in the low corona.

5. Conclusions

The main purpose of this study is to understand the mechanism of the CME flux-rope formation and its relationship with the MC. Notably, we use the Minimum Current Corona model (Longcope, 1996) which, using the pre-flare motions of photospheric magnetic fields and flare-ribbon observations, quantifies the reconnection flux, energy, and helicity budget of the flare. We apply this model to four major eruptive solar flares that produced MCs and compare the predicted flux-rope properties to the observations.

We compare model predictions to observations of four quantities: the predicted model reconnection fluxes to the MC poloidal fluxes and ribbon-motion reconnection fluxes, the predicted flux-rope helicities to the MC helicities, the predicted released energies to the total radiative/conductive energy losses plus the enthalpy fluxes, and the direction of the magnetic field in the AR arcade to the direction of the leading edge of MC poloidal field.

Our comparison reveals the following: The predicted reconnection fluxes match the reconnection fluxes inferred from the observations for the 7 November 2004 and Halloween flares. For the 13 May 2005 and Bastille Day flares, the minimum probable differences between the predicted and observed reconnection fluxes are 13% and 29%, correspondingly. In all four cases the values of poloidal MC fluxes match both the observed and the model reconnection fluxes. The predicted flux-rope helicities match the MC helicities. For three flares of study the predicted free energies lie between the observed energy losses (radiative and conductive energy losses plus the enthalpy fluxes) and the flare luminosities. Only for the flare on 13 May 2005, the predicted free energy is one third of the observed estimate. We relate this mismatch to the fact that 13 May 2005 flare was the only event without a zero-flare, hence additional energy might have been stored before t_0. Finally, we find that in all four cases the direction of the leading MC poloidal field is consistent with the poloidal component of the local AR arcade field, whereas in two cases the MC poloidal-field orientation is opposite to that of the global solar dipole.

These findings compel us to believe that magnetic clouds associated with these four eruptive solar flares are formed by low-corona magnetic reconnection during the eruption, rather than eruption of pre-existing structures in the corona or formation in the upper corona by the global field. Our findings support the conclusions of Qiu *et al.* (2007) and Leamon *et al.* (2004), although through a very different approach: while Qiu *et al.* (2007) and Leamon *et al.* (2004) inferred the solar flux-rope properties only from observations, we infer them from both the MCC model and the observations. Using the pre-flare magnetic field evolution and the MCC model, we find that we are able to predict the observed reconnection fluxes within a 29% uncertainty and the observed MC poloidal-flux and helicity values within the MC length uncertainty. For the flares associated with zero-flares, we are able to estimate a lower limit for the free magnetic energy. We note that, since all four flares occurred in ARs without significant pre-flare flux emergence/cancelation, the flux/energy/helicity that we find is stored by shearing and rotating motions, which is sufficient to account for the observed energy and MC flux and helicity.

Our work brings up several interesting questions that require further exploration. Firstly, the results of this article are based on only a small number of events, which are similar in that all have a large radiative signature. Hence a study of observations of a class of events with small radiative signature would be challenging: smaller flares that are nevertheless associated with major CMEs (Aschwanden, Wuelser, and Nitta, 2009). If the MCC model is valid, it should be able to explain both the energy and helicity content of flare/CME events whose flare energy output is disproportionately small. Secondly, in this article, we estimated the final flare energy as a sum of energy losses by radiation, conduction, and enthalpy, neglecting the energy carried away by the CME. To our knowledge, no systematic empirical study of the source of energy for the CME has yet been conducted, so it is unknown what proportion of its energy budget arises from the AR. From the limited sample, Ravindra and Howard (2010) found that a 50% contribution may be a reasonable first-order approximation. Including CME energy losses into our flare analysis would help us understand how much of the CME energy arises from the active region and may lead to a greater understanding of the onset mechanism for CMEs.

Acknowledgements We thank the TRACE and SOHO/MDI teams for providing the data. SOHO is a project of international cooperation between ESA and NASA. We are pleased to acknowledge support from NASA Earth and Space Science Fellowship grant NNX07AU73H (MDK) and NASA LWS TR&T grant NNG05-GJ96G (RCC and MDK).

Appendix

The X5.7 Bastille Day flare occurred on 14 July 2000 at 10:03 UT in NOAA 9077. Our magnetic-field data describing the evolution of the magnetic field before this flare consist of a sequence of SOHO/MDI full-disk magnetograms (2″, level 1.8) starting at $t_0 = 12$ July 2000 14:27 UT, after the X1.9 flare (12 July 2000 10:18 UT), and ending at $t_{flare} = 14$ July 09:36 UT, 27 minutes before the Bastille day X5.7 flare. Thus we form a sequence of 28 low-noise magnetograms with a 96-minute cadence, which cover 43 hours of the stress build-up prior to the X5.7 flare on 14 July 2000 10:03 UT. Firstly, for all successive pairs of magnetograms we use a Gaussian apodizing window of 7″ to derive a local correlation tracking (LCT) velocity. We then take a magnetogram at t_{flare} and group pixels, exceeding a threshold $B_{thr} = 45$ Gauss downhill from each local maximum, into individual partitions. We combine partitions by eliminating any boundary whose saddle point is less than 350 Gauss below either maximum it separates. Each partition is assigned a unique label which maintains through the sequence by using the LCT velocity pattern. Figure 4 shows the spatial distribution of these partitions at t_{flare}. For expediting the assessment of the field's connectivity, we represent each magnetic partition with a magnetic point charge which contains the magnetic flux of the whole partition concentrated in the partition's centroid. We find that the magnetic field is well balanced at t_{flare} $(\Phi_+(\Phi_-) = 3.3(-3.5) \times 10^{22}$ Mx) and exhibits no significant emergence/cancelation during the 43 hours of pre-flare stress build-up time. From the LCT velocity and magnetic field at each point we find the flux of relative helicity into the corona to be $H_{AR} = -(27 \pm 2.7) \times 10^{42}$ Mx^2, no significant spin-helicity content (rotation) has been detected.

To find the model estimate of the reconnection flux, we determine the magnetic point charges associated with the flare using the flare UV observations by TRACE 1600 Å. Figure 5 shows a superposition of the elements of the topological skeleton at t_{flare} onto the UV flare-ribbon image. The spines (red solid lines) that are associated with ribbons form the footprint of a combination of separatrices that overlay the flaring domains. The overlay suggests that the northern ribbon is associated with the spines connecting flaring point sources P08, P11, P20, P01, P02, P16, P04, and P03; and the southern ribbon is associated with the spines connecting N20, N13, N02, N11, N04, N07, N06, N09, N10, N12, N03, N01, N18, N19, N15, and N05. Field lines connecting the pairs of opposite point charges listed above form a set of flaring domains. The amount of flux that those domains exchanged, the model reconnection flux, is $\Phi_{r,MCC} = (6.0 \pm 0.9) \times 10^{21}$ Mx, fifty percent smaller than the lower value of the observed reconnection flux from the flare-ribbon evolution ($\Phi_{r,ribbon} = (12.8 \pm 3) \times 10^{21}$ Mx).

From the set of flaring point charges and nulls lying between them, we find twenty six flaring separators (see Figure 6). The total free energy and helicity output on those separators is $\mathcal{E}_{MCC} = (9.1 \pm 2.6) \times 10^{31}$ ergs and $H_{MCC} = -(20.1 \pm 3.6) \times 10^{42}$ Mx^2. However, out of the 26 flaring separators, 90% of the total free energy is contained in separators originating in nulls B08 and B11 which lie between poles P01, P02, and P16 (note the most

Figure 4 Positive (P) and negative (N) polarity partitions for NOAA 9077 on 14 July 09:36 UT. The gray-scale magnetogram shows magnetic field scaled from −1000 G to 1000 G. The partitions are outlined and the centroids are denoted by + and × signs (positive and negative, respectively). Axes are labeled in arcseconds from disk center.

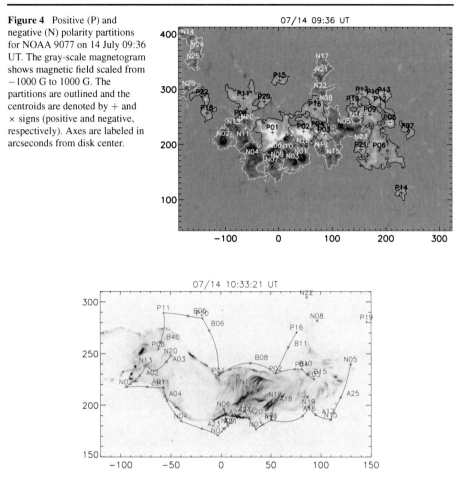

Figure 5 TRACE 1600 Å image, plotted as reverse gray scale, with elements of the topological skeleton superimposed. The skeleton calculated for 14 July 09:36 UT is projected onto the sky after its tangent plane has been rotated to the time of the TRACE observations (10:33 UT). Positive and negative sources are indicated by + and × signs, respectively. The triangles represent the labeled null points. The red curved line segments show spine lines associated with the reconnecting domains. Axes are in arcseconds from disk center.

Figure 6 Flaring separators derived from the MCC model. Colors indicate the logarithm of the free energy available for release during the flare on each separator, in units of ergs.

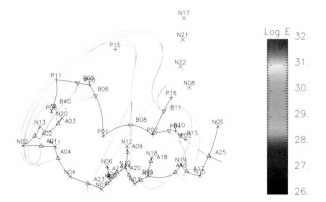

red separators in Figure 6). Moreover, 64% of the total flare free energy is partitioned between three separators: A19/B11, A20/B11, and A20/B08. According to the MCC model, the poles associated with those nulls (P01, P02, P16 and N10, N01, N03, N18) indicate the locations of the largest free-energy release. Figure 6 indicates that the brightest observed loops in TRACE 1600 Å are the loops connecting point charges N12, N10, N01, and N18 with P02, consistent with the results of the MCC model presented above.

References

Abbett, W.P., Fisher, G.H.: 2003, A coupled model for the emergence of active region magnetic flux into the solar corona. *Astrophys. J. Lett.* **582**, L475 – L485. doi:10.1086/344613.

Antiochos, S.K., Devore, C.R., Klimchuk, J.A.: 1999, A model for solar coronal mass ejections. *Astrophys. J. Lett.* **510**, L485 – L493.

Aschwanden, M.J., Wuelser, J.P., Nitta, N.V.: 2009, Solar flares, coronal mass ejections, EUV, stereoscopy. *Solar Phys.* **256**, 3 – 40.

Barnes, G., Longcope, D.W., Leka, K.D.: 2005, Implementing a magnetic charge topology model for solar active regions. *Astrophys. J.* **629**, 561 – 571. doi:10.1086/431175.

Berger, M.A.: 1999, Magnetic helicity in space physics. In: Brown, M.R., Canfield, R.C., Pevtsov, A.A. (eds.) *Magnetic Helicity in Space and Laboratory Plasmas, Geophysical Monograph* **111**, AGU Press, Washington, 1 – 9.

Berger, M.A., Field, G.B.: 1984, The topological properties of magnetic helicity. *J. Fluid Mech.* **147**, 133 – 148. doi:10.1017/S0022112084002019.

Beveridge, C., Longcope, D.W.: 2006, A hierarchical application of the minimum current corona. *Astrophys. J. Lett.* **636**, L453 – L461.

Bothmer, V., Schwenn, R.: 1998, The structure and origin of magnetic clouds in the solar wind. *Ann. Geophys.* **16**, 1 – 24.

Bradshaw, S.J., Cargill, P.J.: 2010, The cooling of coronal plasmas. III. Enthalpy transfer as a mechanism for energy loss. *Astrophys. J.* **717**, 163 – 174. doi:10.1088/0004-637X/717/1/163.

Burlaga, L., Sittler, E., Mariani, F., Schwenn, R.: 1981, Magnetic loop behind an interplanetary shock – voyager, Helios, and imp 8 observations. *J. Geophys. Res.* **86**, 6673 – 6684.

Carmichael, H.: 1964, A process for flares. In: Hess, W.N. (ed.) *AAS-NASA Symposium on the Physics of Solar Flares*, NASA, Washington, 451.

Chae, J.: 2001, Observational determination of the rate of magnetic helicity transport through the solar surface via the horizontal motion of field line footpoints. *Astrophys. J. Lett.* **560**, L95 – L98. doi:10.1086/324173.

Chae, J.: 2007, Measurements of magnetic helicity injected through the solar photosphere. *Adv. Space Res.* **39**, 1700 – 1705. doi:10.1016/j.asr.2007.01.035.

Chen, J.: 1989, Effects of toroidal forces in current loops embedded in a background plasma. *Astrophys. J. Lett.* **338**, L453 – L470.

Crooker, N.U.: 2000, Solar and heliospheric geoeffective disturbances. *J. Atmos. Solar-Terr. Phys.* **62**, 1071 – 1085.

Dasso, S., Mandrini, C.H., Démoulin, P., Farrugia, C.J.: 2003, Magnetic helicity analysis of an interplanetary twisted flux tube. *J. Geophys. Res.* **108**, 3 – 1. doi:10.1029/2003JA009942.

Dasso, S., Mandrini, C.H., Démoulin, P., Luoni, M.L.: 2006, A new model-independent method to compute magnetic helicity in magnetic clouds. *Astron. Astrophys.* **455**, 349 – 359. doi:10.1051/0004-6361:20064806.

Démoulin, P.: 2008, A review of the quantitative links between CMEs and magnetic clouds. *Ann. Geophys.* **26**, 3113 – 3125.

Démoulin, P., Pariat, E.: 2007, Computing magnetic energy and helicity fluxes from series of magnetograms. *Mem. Soc. Astron. Italiana* **78**, 136.

Démoulin, P., Mandrini, C.H., van Driel-Gesztelyi, L., Thompson, B.J., Plunkett, S., Kovári, Z., Aulanier, G., Young, A.: 2002, What is the source of the magnetic helicity shed by CMEs? The long-term helicity budget of AR 7978. *Astron. Astrophys.* **382**, 650 – 665. doi:10.1051/0004-6361:20011634.

DeVore, C.R.: 2000, Magnetic helicity generation by solar differential rotation. *Astrophys. J. Lett.* **539**, L944 – L953.

Fan, Y., Gibson, S.E.: 2004, Numerical simulations of three-dimensional coronal magnetic fields resulting from the emergence of twisted magnetic flux tubes. *Astrophys. J. Lett.* **609**, L1123 – L1133. doi:10.1086/421238.

Fletcher, L., Hudson, H.: 2001, The magnetic structure and generation of EuV flare ribbons. *Solar Phys.* **204**, 69–89.

Forbes, T.G., Priest, E.R.: 1995, Photospheric magnetic field evolution and eruptive flares. *Astrophys. J.* **446**, 377. doi:10.1086/175797.

Gibson, S.E., Fan, Y.: 2008, Partially ejected flux ropes: implications for interplanetary coronal mass ejections. *J. Geophys. Res.* **113**, 9103. doi:10.1029/2008JA013151.

Gopalswamy, N., Yashiro, S., Kaiser, M.L., Howard, R.A., Bougeret, J.: 2001, Radio signatures of coronal mass ejection interaction: coronal mass ejection cannibalism? *Astrophys. J. Lett.* **548**, L91–L94. doi:10.1086/318939.

Gopalswamy, N., Akiyama, S., Yashiro, S., Mäkelä, P.: 2010, Coronal mass ejections from sunspot and nonsunspot regions. In: Hasan, S.S. and Rutten, R.J. (eds.) *Magnetic Coupling Between the Interior and Atmosphere of the Sun, Astrophys. Space Science Proc.*, Springer, Berlin, 289–307. doi:10.1007/978.

Gosling, J.T.: 1990, Coronal mass ejections and magnetic flux ropes in interplanetary space. In: Russel, C.T., Priest, E.R., Lee, L.C. (eds.) *Physics of Magnetic Flux Ropes, Geophys. Monographs* **58**, AGU, Washington, 343–364.

Green, L.M., López Fuentes, M.C., Mandrini, C.H., Démoulin, P., Van Driel-Gesztelyi, L., Culhane, J.L.: 2002, The magnetic helicity budget of a CME-prolific active region. *Solar Phys.* **208**, 43–68.

Gulisano, A.M., Dasso, S., Mandrini, C.H., Démoulin, P.: 2005, Magnetic clouds: a statistical study of magnetic helicity. *J. Atmos. Solar-Terr. Phys.* **67**, 1761–1766. doi:10.1016/j.jastp.2005.02.026.

Hirayama, T.: 1974, Theoretical model of flares and prominences. i: Evaporating flare model. *Solar Phys.* **34**, 323–338.

Hu, Q., Sonnerup, B.U.Ö.: 2001, Reconstruction of magnetic flux ropes in the solar wind. *Geophys. Res. Lett.* **28**, 467–470.

Jing, J., Lee, J., Liu, C., Gary, D.E., Wang, H.: 2007, Hard X-ray intensity distribution along Hα ribbons. *Astrophys. J. Lett.* **664**, L127–L130. doi:10.1086/520812.

Kazachenko, M.D., Canfield, R.C., Longcope, D.W., Qiu, J., Des Jardins, A., Nightingale, R.W.: 2009, Sunspot rotation, flare energetics, and flux rope helicity: the eruptive flare on 2005 May 13. *Astrophys. J.* **704**, 1146–1158. doi:10.1088/0004-637X/704/2/1146.

Kazachenko, M.D., Canfield, R.C., Longcope, D.W., Qiu, J.: 2010, Sunspot rotation, flare energetics, and flux rope helicity: the Halloween flare on 2003 28 October. *Astrophys. J.* **722**, 1539–1546. doi:10.1088/0004-637X/722/2/1539.

Kopp, R.A., Pneuman, G.W.: 1976, Magnetic reconnection in the corona and the loop prominence phenomenon. *Solar Phys.* **50**, 85–98.

Larson, D.E., Lin, R.P., McTiernan, J.M., McFadden, J.P., Ergun, R.E., McCarthy, M., Rème, H., Sanderson, T.R., Kaiser, M., Lepping, R.P., Mazur, J.: 1997, Tracing the topology of the October 18–20, 1995, magnetic cloud with $0.1 - 10^2$ keV electrons. *Geophys. Res. Lett.* **24**, 1911–1914.

Leamon, R.J., Canfield, R.C., Pevtsov, A.A.: 2002, Properties of magnetic clouds and geomagnetic storms associated eruption of coronal sigmoids. *J. Geophys. Res.* **107**, 1234.

Leamon, R.J., Canfield, R.C., Jones, S.L., Lambkin, K., Lundberg, B.J., Pevtsov, A.A.: 2004, Helicity of magnetic clouds and their associated active regions. *J. Geophys. Res.* **109**, 5106. doi:10.1029/2003JA010324.

Leka, K.D., Canfield, R.C., McClymont, A.N., Van Driel Gesztelyi, L.: 1996, Evidence for current-carrying emerging flux. *Astrophys. J. Lett.* **462**, L547–L560.

Lepping, R.P., Berdichevsky, D.B., Burlaga, L.F., Lazarus, A.J., Kasper, J., Desch, M.D., Wu, C., Reames, D.V., Singer, H.J., Smith, C.W., Ackerson, K.L.: 2001, The Bastille day magnetic clouds and upstream shocks: near-earth interplanetary observations. *Solar Phys.* **204**, 285–303. doi:10.1023/A:1014264327855.

Li, Y., Luhman, J.G., Lynch, B.J., Kilpua, E.: 2011, Cyclic reversal of magnetic cloud poloidal field. *Solar Phys.* **269**, 32–47. doi:10.1007/s11207-011-9722-9.

Lin, J., Raymond, J.C., van Ballegooijen, A.A.: 2004, The role of magnetic reconnection in the observable features of solar eruptions. *Astrophys. J.* **602**, 422–435. doi:10.1086/380900.

Liu, C., Lee, J., Yurchyshyn, V., Deng, N., Cho, K.s., Karlický, M., Wang, H.: 2007, The eruption from a sigmoidal solar active region on 2005 May 13. *Astrophys. J.* **669**, 1372–1381. doi:10.1086/521644.

Liu, J., Zhang, Y., Zhang, H.: 2008, Relationship between powerful flares and dynamic evolution of the magnetic field at the solar surface. *Solar Phys.* **248**, 67–84. doi:10.1007/s11207-008-9149-0.

Longcope, D.W.: 1996, Topology and current ribbons: a model for current, reconnection and flaring in a complex, evolving corona. *Solar Phys.* **169**, 91–121. doi:10.1007/BF00153836.

Longcope, D.W.: 2001, Separator current sheets: generic features in minimum-energy magnetic fields subject to flux constraints. *Phys. Plasmas* **8**, 5277–5290. doi:10.1063/1.1418431.

Longcope, D.W., Magara, T.: 2004, A comparison of the minimum current corona to a magnetohydrodynamic simulation of quasi-static coronal evolution. *Astrophys. J.* **608**, 1106–1123. doi:10.1086/420780.

Longcope, D.W., DesJardins, A.C., Carranza-Fulmer, T., Qiu, J.: 2010, A quantitative model of energy release and heating by time-dependent, localized reconnection in a flare with a thermal loop-top X-ray source. *Solar Phys.* **267**, 107 – 139. doi:10.1007/s11207-010-9635.

Longcope, D.W., Beveridge, C., Qiu, J., Ravindra, B., Barnes, G., Dasso, S.: 2007, Modeling and measuring the flux reconnected and ejected by the two-ribbon flare/CME event on 7 November 2004. *Solar Phys.* **244**, 45 – 73. doi:10.1007/s11207-007-0330-7.

Longcope, D.W., Barnes, G., Beveridge, C.: 2009, Effects of partitioning and extrapolation on the connectivity of potential magnetic fields. *Astrophys. J.* **693**, 97 – 111. doi:10.1088/0004-637X/693/1/97.

Low, B.C.: 1994, Magnetohydrodynamic processes in the solar corona: flares, coronal mass ejections, and magnetic helicity. *Phys. Plasmas* **1**, 1684 – 1690.

Luoni, M.L., Mandrini, C.H., Dasso, S., van Driel-Gesztelyi, L., Démoulin, P.: 2005, Tracing magnetic helicity from the solar corona to the interplanetary space. *J. Atmos. Solar-Terr. Phys.* **67**, 1734 – 1743. doi:10.1016/j.jastp.2005.07.003.

Lynch, B.J., Gruesbeck, J.R., Zurbuchen, T.H., Antiochos, S.K.: 2005, Solar cycle-dependent helicity transport by magnetic clouds. *J. Geophys. Res.* **110**, 8107. doi:10.1029/2005JA011137.

Mackay, D.H., van Ballegooijen, A.A.: 2006, Models of the large-scale corona. I. Formation, evolution, and liftoff of magnetic flux ropes. *Astrophys. J.* **641**, 577 – 589. doi:10.1086/500425.

Mandrini, C.H., Pohjolainen, S., Dasso, S., Green, L.M., Démoulin, P., van Driel-Gesztelyi, L., Copperwheat, C., Foley, C.: 2005, Interplanetary flux rope ejected from an X-ray bright point. The smallest magnetic cloud source-region ever observed. *Astron. Astrophys.* **434**, 725 – 740. doi:10.1051/0004-6361: 20041079.

Mandrini, C.H., Demoulin, P., Schmieder, B., Deluca, E.E., Pariat, E., Uddin, W.: 2006, Companion event and precursor of the X17 flare on 28 October 2003. *Solar Phys.* **238**, 293 – 312. doi:10.1007/s11207-006-0205-3.

Marubashi, K.: 1986, Structure of the interplanetary magnetic clouds and their solar origins. *Adv. Space Res.* **6**, 335 – 338. doi:10.1016/0273-1177(86)90172-9.

Masuda, S., Kosugi, T., Hudson, H.S.: 2001, A hard X-ray two-ribbon flare observed with Yohkoh/HXT. *Solar Phys.* **204**, 55 – 67. doi:10.1023/A:1014230629731.

Mewe, R., Gronenschild, E.H.B.M., van den Oord, G.H.J.: 1985, Calculated X-radiation from optically thin plasmas. V. *Astron. Astrophys. Suppl. Ser.* **62**, 197 – 254.

Mulligan, T., Russell, C.T., Luhmann, J.G.: 1998, Solar cycle evolution of the structure of magnetic clouds in the inner heliosphere. *Geophys. Res. Lett.* **25**, 2959 – 2962. doi:10.1029/98GL01302.

Nindos, A., Zhang, J., Zhang, H.: 2003, The magnetic helicity budget of solar active regions and coronal mass ejections. *Astrophys. J. Lett.* **594**, 1033 – 1048. doi:10.1086/377126.

November, L.J., Simon, G.W.: 1988, Precise proper-motion measurement of solar granulation. *Astrophys. J.* **333**, 427 – 442. doi:10.1086/166758.

Pariat, E., Démoulin, P., Berger, M.A.: 2005, Photospheric flux density of magnetic helicity. *Astron. Astrophys.* **439**, 1191 – 1203.

Pevtsov, A.A., Balasubramaniam, K.S.: 2003, Helicity patterns on the sun. *Adv. Space Res.* **32**, 1867 – 1874. doi:10.1016/S0273-1177(03)90620-X.

Poletto, G., Kopp, R.A.: 1986, Macroscopic electric fields during two-ribbon flares. In: Neidig, D.F. (ed.) *The Lower Atmospheres of Solar Flares*, National Solar Observatory, Sunspot, 453 – 465.

Qiu, J., Gary, D.E.: 2003, Flare-related magnetic anomaly with a sign reversal. *Astrophys. J.* **599**, 615 – 625. doi:10.1086/379146.

Qiu, J., Hu, Q., Howard, T.A., Yurchyshyn, V.B.: 2007, On the magnetic flux budget in low-corona magnetic reconnection and interplanetary coronal mass ejections. *Astrophys. J.* **659**, 758 – 772. doi:10.1086/512060.

Ravindra, B., Howard, T.A.: 2010, Comparison of energies between eruptive phenomena and magnetic field in AR 10930. *Bull. Astron. Soc. India* **38**, 147 – 163. http://cdsads.u-strasbg.fr/abs/2010BASI...38..147R.

Régnier, S., Priest, E.R.: 2007, Free magnetic energy in solar active regions above the minimum-energy relaxed state. *Astrophys. J. Lett.* **669**, L53 – L56. doi:10.1086/523269.

Rosner, R., Tucker, W.H., Vaiana, G.S.: 1978, Dynamics of the quiescent corona. *Astrophys. J. Lett.* **220**, 643 – 655.

Scherrer, P.H., Bogart, R.S., Bush, R.I., Hoeksema, J.T., Kosovichev, A.G., Schou, J., Rosenberg, W., Springer, L., Tarbell, T.D., Title, A., Wolfson, C.J., Zayer, I., MDI Engineering Team: 1995, The solar oscillations investigation – Michelson Doppler imager. *Solar Phys.* **162**, 129 – 188. doi:10.1007/BF00733429.

Schrijver, C.J., Derosa, M.L., Metcalf, T.R., Liu, Y., McTiernan, J., Régnier, S., Valori, G., Wheatland, M.S., Wiegelmann, T.: 2006, Nonlinear force-free modeling of coronal magnetic fields Part I: A quantitative comparison of methods. *Solar Phys.* **235**, 161 – 190. doi:10.1007/s11207-006-0068-7.

Sturrock, P.A.: 1968, A model of solar flares. In: Kiepenheuer, K.O. (ed.) *IAU Symp. 35: Structure and Development of Solar Active Regions*, *IAU Symp.* **35**, Reidel, Dordrecht, 471 – 479.

Su, Y., Golub, L., Van Ballegooijen, A.A.: 2007, A statistical study of shear motion of the footpoints in two-ribbon flares. *Astrophys. J.* **655**, 606 – 614. doi:10.1086/510065.

Vesecky, J.F., Antiochos, S.K., Underwood, J.H.: 1979, Numerical modeling of quasi-static loops. I. Uniform energy input. *Astrophys. J. Lett.* **233**, L987 – L997.

Webb, D.F., Lepping, R.P., Burlaga, L.F., Deforest, C.E., Larson, D.E., Martin, S.F., Plunkett, S.P., Rust, D.M.: 2000, The origin and development of the May 1997 magnetic cloud. *J. Geophys. Res.* **105**, 27251 – 27260.

Welsch, B.T., Abbett, W.P., DeRosa, M.L., Fisher, G.H., Georgoulis, M.K., Kusano, K., Longcope, D.W., Ravindra, B., Schuck, P.W.: 2007, Tests and comparisons of velocity inversion techniques. *Astrophys. J. Lett.* **670**, L1434 – L1452.

Woods, T.N., Kopp, G., Chamberlin, P.C.: 2006, Contributions of the solar ultraviolet irradiance to the total solar irradiance during large flares. *J. Geophys. Res.* **111**, 10. doi:10.1029/2005JA011507.

Yurchyshyn, V., Abramenko, V., Tripathi, D.: 2009, Rotation of white-light coronal mass ejection structures as inferred from LASCO coronagraph. *Astrophys. J.* **705**, 426 – 435. doi:10.1088/0004-637X/705/1/426.

Yurchyshyn, V., Hu, Q., Abramenko, V.: 2005, Structure of magnetic fields in NOAA active regions 0486 and 0501 and in the associated interplanetary ejecta. *Space Weather* **3**, 8. doi:10.1029/2004SW000124.

Yurchyshyn, V.B., Wang, H., Goode, P.R., Deng, Y.: 2001, Orientation of the magnetic fields in interplanetary flux ropes and solar filaments. *Astrophys. J.* **563**, 381 – 388. doi:10.1086/323778.

Yurchyshyn, V., Liu, C., Abramenko, V., Krall, J.: 2006, The May 13, 2005 eruption: observations, data analysis and interpretation. *Solar Phys.* **239**, 317 – 335. doi:10.1007/s11207-006-0177-3.

Yurchyshyn, V., Hu, Q., Lepping, R.P., Lynch, B.J., Krall, J.: 2007, Orientations of LASCO Halo CMEs and their connection to the flux rope structure of interplanetary CMEs. *Adv. Space Res.* **40**, 1821 – 1826. doi:10.1016/j.asr.2007.01.059.

Zhang, Y., Liu, J., Zhang, H.: 2008, Relationship between rotating sunspots and flares. *Solar Phys.* **247**, 39 – 52. doi:10.1007/s11207-007-9089-0.

Zhao, X.P., Hoeksema, J.T.: 1998, Central axial field direction in magnetic clouds and its relation to southward interplanetary magnetic field events and dependence on disappearing solar filaments. *J. Geophys. Res.* **103**, 2077 – 2083. doi:10.1029/97JA03234.

Solar Phys (2012) 277:185–201
DOI 10.1007/s11207-011-9881-8

Coronal Mass Ejections from Magnetic Systems Encompassing Filament Channels Without Filaments

Alexei A. Pevtsov · Olga Panasenco · Sara F. Martin

Received: 13 January 2011 / Accepted: 17 October 2011 / Published online: 15 November 2011
© Springer Science+Business Media B.V. 2011

Abstract Well-developed filament channels may be present in the solar atmosphere even when there is no trace of filament material inside them. Such magnetic systems with filament channels without filaments can result in coronal mass ejections that might appear to have no corresponding solar surface source regions. In this case study, we analyze CMEs on 9 August 2001 and 3 March 2011 and trace their origins to magnetic systems with filament channels containing no obvious filament material on the days around the eruptions.

Keywords Coronal mass ejections, low coronal signatures · Coronal mass ejections, initiation and propagation · Magnetic fields, corona · Prominences, formation and evolution · Filaments, filament mass

1. Introduction

Coronal mass ejections (CMEs) are typically associated with flares or filament eruptions. Some studies suggest that the majority of CMEs can be traced back to filament eruptions (Subramanian and Dere, 2001). There are, however, cases of CME when no flare or filament eruption has been observed. For example, Robbrecht, Patsourakos, and Vourlidas (2009) reported a STEREO observation of a CME without a clear signature of a solar source region in the photosphere, chromosphere, or low corona. A recent statistical study by Ma *et al.* (2010) finds that about one third of CMEs observed in 2009 exhibit no signature of the eruption

Solar Flare Magnetic Fields and Plasmas
Guest Editors: Y. Fan and G.H. Fisher

A.A. Pevtsov (✉)
National Solar Observatory, Sunspot, NM 88349, USA
e-mail: apevtsov@nso.edu

O. Panasenco · S.F. Martin
Helio Research, La Crescenta, CA 91214, USA

O. Panasenco
e-mail: OlgaPanasenco@aol.com

in the low corona. We suggest that some of these CMEs "from nowhere" may originate in magnetic systems encompassing filament channels without conspicuous filaments.

A filament channel is a necessary "feature" of the chromospheric-coronal filament system (for review, see Gaizauskas, 1998). Filament channels overlay magnetic polarity reversal boundaries (also called polarity inversion lines or "neutral lines"). However, a filament channel is not a line, but a volume of space around and encompassing a filament (when it is present) or its future location. Not every polarity reversal boundary has a corresponding filament channel above and around it: a vital precondition for the formation of a filament channel is a magnetic field in it having a strong horizontal component aligned with the polarity reversal boundary (Gaizauskas, Mackay, and Harvey, 2001). In Hα, filament channels can be distinguished by "voids" of coronal mass and an anti-parallel pattern of fibrils on opposite sides of a polarity reversal boundary (Foukal, 1971; Martin, 1998; Martin, Lin, and Engvold, 2008; Martin and Panasenco, 2010; Panasenco, 2010). As summarized by Martin (1998) no fibrils cross the polarity reversal line in a fully developed filament channel. Because fibrils are field-aligned this implies that the same is probably true for magnetic field lines associated with fibrils, *i.e.*, no magnetic field lines from active region or network magnetic fields cross this polarity reversal boundary at the chromospheric level (Smith, 1968; Martin, 1990). In other words, magnetic systems associated with filament channels do not have low-lying long-lived loops across the channel except in their flaring or post-flaring state (Martin, 1990). However, long-lived loops always exist high above the channel and high above filament mass in a channel (Martin, 1990).

Filament channels are readily identifiable in He II 304 Å images during the active phase of solar cycles as dark narrow "corridors", but they are not clearly visible during the minimum of solar cycles or in circumstances when there is very little magnetic flux surrounding the polarity reversal boundary. Similarly, dark filament channels can be identified in 195 Å images (*e.g.*, Vásquez, Frazin, and Kamalabadi, 2009). Apparent coronal voids often coincide with filament channels. Not every polarity reversal boundary has a filament channel, and not every filament channel has a filament (Gaizauskas, 1998). Panasenco and Pevtsov (2010) have analyzed the evolution of a stable and apparently empty filament channel during a period of low solar activity, and found that it had some of the ingredients necessary for filament formation:

i) A magnetic neutral line.
ii) An arcade field overlying the neutral line (Martin, 1990).
iii) A coronal cavity below the arcade.

Panasenco and Pevtsov (2010) did observe an episode, when relatively hot chromospheric material (as observed in He II 304 Å) was injected into the filament channel from outside (see, Figure 1 in Panasenco and Pevtsov, 2010). However, the emission from this material rapidly disappeared as the plasma cooled down. Based on the analysis of magnetic fields, Panasenco and Pevtsov (2010) concluded that the prime reason for filaments not forming in that channel was a deficiency in the mechanism that supplies mass for filaments. Several studies have indicated that the existence of filaments is closely related to the magnetic flux cancellation rate within the filament channel (*e.g.*, Martin, Livi, and Wang, 1985; Martin, 1990; Litvinenko, 1999; Litvinenko and Martin, 1999; Wood and Martens, 2003). In a more recent study, Mackay, Gaizauskas, and Yeates (2008) confirmed that the convergence of magnetic flux leading to either flux cancellation or reconnection is required for filament formation.

Partially empty or completely empty filament channels can be observed not only in the quiet Sun, but also in active regions (Zirker *et al.*, 1997; Martin, 1998). Filament channels

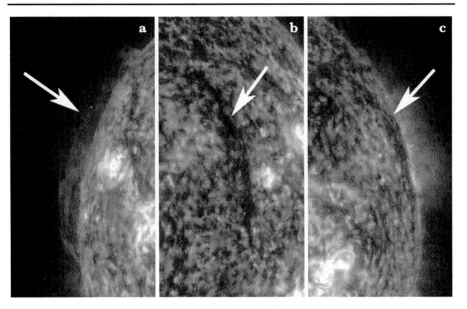

Figure 1 SOHO/EIT 304 Å images on (a) 01 August 2001 at 19:19 UT, (b) 11 August 2001 at 07:19 UT, (c) 14 August 2001 at 07:19 UT. The prominence at the east limb (a) corresponds to the filament channel observed as a dark void in 304 Å and indicated by the arrows at (b) and (c).

without filament mass in Hα in active regions are less common than among the more dispersed and lower density magnetic fields on the quiet Sun. In this paper, we present two case studies of CME originating from a magnetic system on the quiet Sun; each encompasses a filament channel largely devoid of filament mass during two to four days prior to the eruption of the filament. Informally the term "empty" has been occasionally used to refer to filament channels without the presence of a conspicuous filament material as observed in Hα and/or He II 304 Å. We cannot rule out that there could be some material present with a low-enough column density that is undetectable by observations. For example, Heinzel *et al.* (2008) have found a column density in Hα filaments in a range of $1-5 \times 10^{-5}$ g cm^{-2}. The results of their investigation suggest that if the column density is below 1×10^{-5} g cm^{-2}, no filament will be seen in Hα observations.

2. Magnetic Systems Encompassing Empty Filament Channels

2.1. Case 1: Eruption from Empty Filament Channel on 9 August 2001

Our first case of a magnetic system with an empty filament channel was observed throughout its entire disk passage during 1 – 14 August 2001. Hα observations show no indication of continuous filament material in the observed channel during the entire period of observations. He II 304 Å data also show no filament except on 1 and 2 August when a relatively hot material visible in 304 Å was periodically injected into the filament channel from one of its ends. Images taken during these two days show a prominence forming along the filament channel (see Figure 1a). However, this prominence/filament does not persist. Its material cools down rapidly and vanishes. This filament plasma behavior is similar to one described in Panasenco and Pevtsov (2010) and can be evidence of a filament channel with periodic flow of mass along the filament channel without a clear Hα or 304 Å filament in it. As the region rotates onto the disk, the filament channel becomes visible as a dark void in EUV lines (Figure 1b, c). There are also small fragments of filament mass visible in Hα that can be identified in a few places along the filament channel. Similar mass fragments were described in Pevtsov and Neidig (2005). They might correspond to filament pillars (barbs) extending down to the chromosphere. In Hα images, higher density exists in the barbs and lower parts of filaments whereas in He II 304 Å images, the upper parts of the filament are brighter or not seen in Hα (Lin, 2004).

Figure 2a, b shows an Hα image of the solar disk and a corresponding MDI magnetogram on 6 August 2001. The white box marks the approximate location of the filament channel without a filament, and the corresponding area of the magnetic polarity reversal boundary in the photosphere. The channel has an approximate length of 50 solar degrees. It runs nearly along the solar meridian and extends from the northern to the southern hemisphere across the equator. The extrapolation of the photospheric magnetic field using a Potential Field Source Surface (PFSS) model shows the coronal arcade above the filament channel (Figure 2c). Here and in the following discussion we use the PFSS model to represent a possible magnetic connectivity in the corona. Proper modeling of the magnetic environment in the filament channel would require a more sophisticated model, and is outside the scope of this paper. From analyzing the data we conclude that this channel developed during four previous solar rotations. We estimate that at least five decaying regions from both hemispheres contributed to its formation. An example of such long-lived development of a filament channel has been described by Gaizauskas, Mackay, and Harvey (2001).

Figure 3 shows the evolution of the area marked on Figure 2a during its disk passage (to save space, data for 7 August are not shown). The first three images in Figure 3a – c (5, 6 and 8 August) show the filament channel before its eruption. The small fragments of the filament plasma are still visible in Hα on 5 and 6 August but completely disappear two days before the eruption on 9 August 2001. One can speculate that such a complete disappearance might be due to a slow rise of the magnetic system of the filament channel and the surrounding coronal loop system before its eruption accompanied by an expansion of the filament volume and a corresponding decrease in the filament density; alternatively the disappearance might be due to the quenching of the mechanism supplying filament material to the channel. As described in Pevtsov and Neidig (2005), the fragments of dark material tracing the filament channel as observed in Hα represent the filament barbs and their disappearance may be due to gradual detachment from the chromosphere during slow rise phase of the filament system before its eruption. Figure 3d shows the filament channel on 9 August, approximately seven hours after the beginning of the eruption. Figure 3e – g show partial reformation of some filament fragments after the eruption.

The filament channel just before the associated CME eruption is shown in Figure 4 (top). The void of the filament cavity observed against the disk in 195 Å is indicated by white arrows. Flare-like ribbons at the ends of the loops are rendered more visible by wavelet image processing (Figure 4, bottom). The direction of the skew of the flare loops is left-handed, which when combined with known one-to-one chiral relationships (Martin, 1998) implies dextral chirality for this filament channel. In Figure 5 the post-eruption trans-equatorial arcade is observed in the EIT 284 Å image.

A partial halo coronal mass ejection was observed in the LASCO C2 coronagraph on 9 August 2001 by 10:54 UT above NW limb (Figure 6). At first glance, coronal activity on the disk did not show any dramatic evolution suggesting the occurrence of a CME. Upon close examination, however, the changes in the corona around the long trans-equatorial filament channel provide evidence for this channel to be the key site around which the CME occurs.

Figure 2 (a) The filament channel of the prominence in Figure 1, here marked by a white box. It was recorded in Hα at BBSO, 06 August 2001 at 16:54 UT; (b) MDI magnetogram and (c) PFSS extrapolation of the magnetic arcade above the filament channel on 06 August 2001 at 16:03 UT. While the PFSS model is an approximation of the coronal field lines, it confirms the continued existence of the channel where it appears to be empty; an overlying coronal arcade which is one of the necessary conditions for the existence of a filament channel (Martin, 1990).

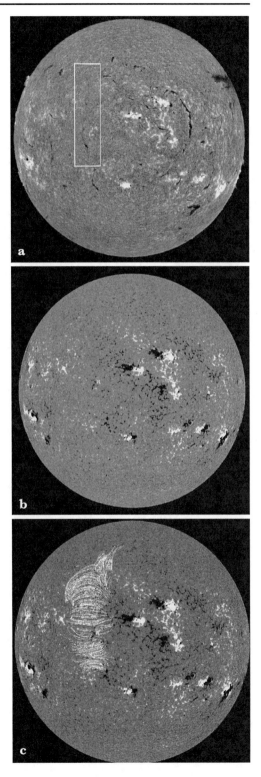

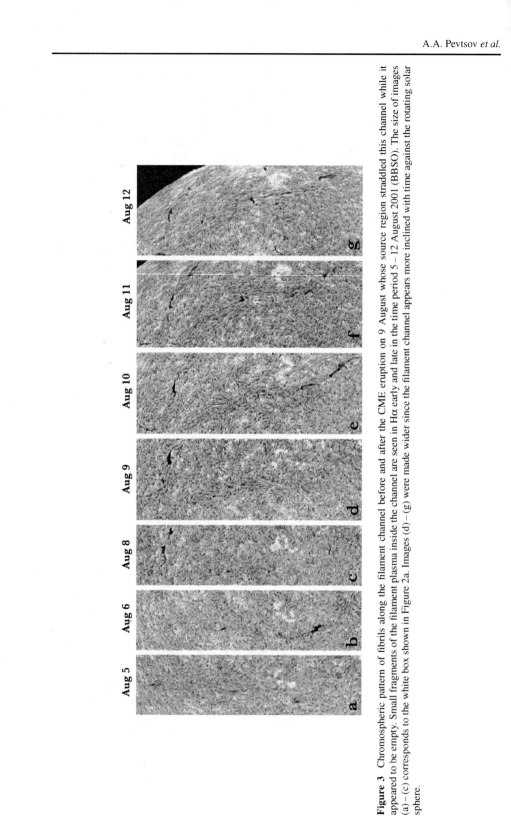

Figure 3 Chromospheric pattern of fibrils along the filament channel before and after the CME eruption on 9 August whose source region straddled this channel while it appeared to be empty. Small fragments of the filament plasma inside the channel are seen in Hα early and late in the time period 5 – 12 August 2001 (BBSO). The size of images (a) – (c) corresponds to the white box shown in Figure 2a. Images (d) – (g) were made wider since the filament channel appears more inclined with time against the rotating solar sphere.

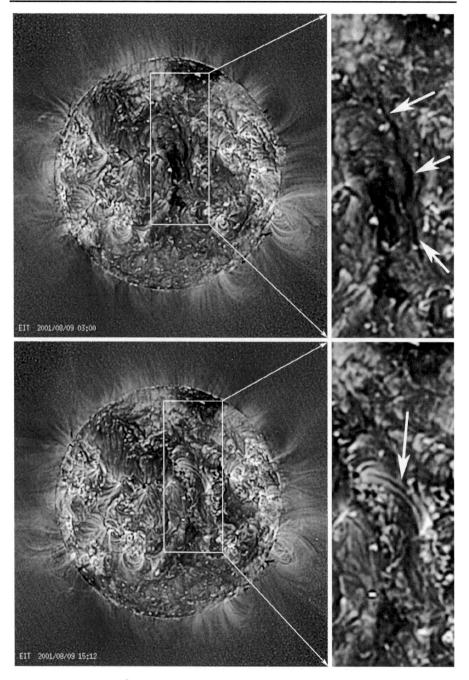

Figure 4 SOHO/EIT 195 Å wavelet-enhanced images before (top row) and after the eruption (bottom row) around 10:54 UT on 9 August. Upper row: at the left is shown the filament channel before the eruption inside a white box; at the right, white arrows point to the 195 Å dark linear feature identifying the filament channel (09 August 2001 at 03:00 UT). Bottom row: filament channel soon after the eruption; A white arrow points to flare-loop like features above the filament channel. The left-skew of these loops is evidence that the filament channel beneath has dextral chirality according to the established one-to-one solar chiral relationships. (09 August 2001 at 15:12 UT.)

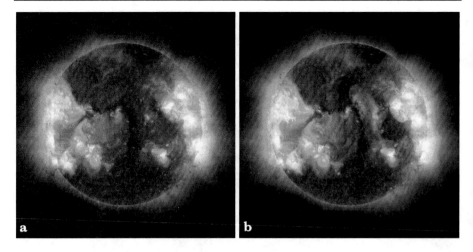

Figure 5 SOHO/EIT 284 Å images taken prior to a CME (a: 9 August 2001, 7:06 UT) and early in the eruption (b: 9 August 2001, 12:06 UT). New, post-eruption, trans-equatorial, bright, unresolved structure is clearly visible near west of an isolated coronal hole that is elongated and has a north–south orientation and crosses the equator near central meridian.

Figure 6 Although originating from near disk center, this partial halo CME is seen entirely at the west limb. This is evidence that the CME was strongly non-radial away from the elongated coronal hole nearly along the central meridian in Figure 5. (SOHO/LASCO C2 09 August 2001 at 12:30 UT.)

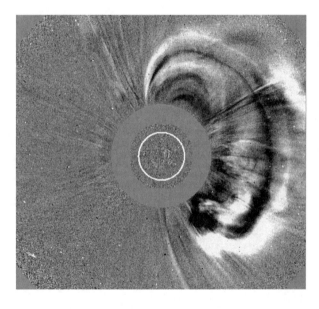

The timing of events in this first case, observed on the solar disk in EIT 195 Å data, provides observational evidence for a CME originating from above and around the trans-equatorial filament channel situated near the disk center on 9 August. No filament material was observed in this channel in Hα or He II 304 Å spectral lines. Therefore, we refer to it as an "empty" filament channel at the time of the associated CME and during the preceding day. A wide post-eruption trans-equatorial set of "loops" crossing the filament channel was observed in EIT 284 Å (Figure 5b) and soft X-ray (*Yohkoh*) images, which indicates the presence of high temperature coronal plasma formed at the sides of the filament channel during the CME formation. Hα images show no brightenings along the filament channel that

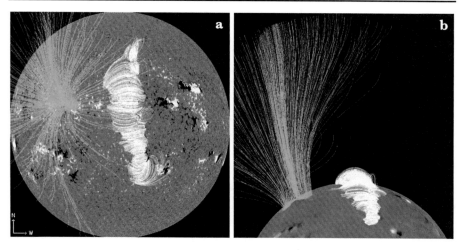

Figure 7 Magnetic field lines from the PFSS model on 08 August 2001 18:04 UT (a) "on disk" and (b) "off limb" projections. White lines correspond to closed field lines of the arcade overlying the filament channel; the green lines are open field lines originating in an isolated coronal hole east of the filament channel. Under these circumstances, any CME originating from the coronal arcade is expected to be non-radial to the west away from the magnetic field of the coronal hole as indicated in Figure 6.

could indicate a reconnection process if a "flux rope" structure was formed at the beginning of the eruption. On the other hand, we do see two compact brightenings near the northern and southern ends of the filament channel in the early stages of the eruption (no figure is shown). These brightenings suggest that a horizontal structure with at least one continuous field line connecting the two ends of the filament channel has already been present at the very early stages of the eruption. Additionally, EIT 304 Å and 195 Å images show weak brightenings on opposite sides of the filament channel near its middle part, as well as thin bright loops reminiscent of post-flare loops. This evolution observed during the early stages of the eruption around this filament channel shows several aspects typical for the classical two-ribbon flare scenario, albeit with much weaker intensity.

The potential field extrapolation (Figures 2c and 7) reveals the presence of a magnetic arcade above the filament channel. The magnetic polarity of the footpoints on the eastern side of the arcade is positive, the same as that of the coronal hole situated to the east of the filament channel. In the LASCO C2 coronagraph, the CME front can be clearly identified in images taken at 10:54 UT. Using difference images, we manually traced the CME's front in five consecutive images. The change in height of the CME front (F_{CME}) with time (t) was fitted by second degree polynomial to arrive at an initial speed of 257 km s^{-1} and an acceleration of ≈ 0.024 km s^{-2} relative to its first appearance at 10:54 UT: $F_{CME} = (257 \pm 20) \cdot t + (0.012 \pm 0.002) \cdot t^2$. By the time the CME left the LASCO C2 field of view, the CME front was moving with a speed of about $450 - 480$ km s^{-1} in the image plane.

Although the CME had originated around the filament channel near the disk center, the CME was observed propagating mostly from the west limb of the Sun. We suggest, based on use of the PFSS modeling (Figure 7), that the CME had non-radial motion that is consistent with deflection toward the west by the open magnetic field of an isolated coronal hole. Deflection of CMEs by coronal hole boundaries has been reported by several researchers (*e.g.*, Gopalswamy *et al.*, 2003, 2009; Cremades and Bothmer, 2004; Panasenco *et al.*, 2011). Figure 7 shows that the approximate configuration of the coronal hole, situated farther away from the filament channel, would allow deflection of the CME to take place high in the corona.

Between 12 August (about 10:40 UT) and 13 August ($\approx$ 11:00 UT), *in situ* measurements from ACE and WIND spacecraft registered a passage of an interplanetary CME (ICME). Taking the first indication of the ICME disturbance in ACE data as the ICME front yields an average ICME velocity of about 580 km s^{-1}, which seems to be in general agreement with estimates based on LASCO observations. The ICME had a complicated structure with multiple reversals in the direction of all three components of its magnetic field. Due to this complexity, we did not attempt fitting any axi-symmetric model of magnetic cloud to ACE data.

Geomagnetic measurements of the A$_p$-index show an onset of a major geomagnetic storm at approximately 22:40 UT on 12 August (A$_p$-index = 56 at storm's maximum at 4:30:43 UT on 13 August). Using the calculated time difference ($\approx$ 3.5 days) between the estimated CME eruption and the beginning of geomagnetic storm yields an average speed of the CME of about 500 km s^{-1}, which, again, is in agreement with CME speed measured from LASCO C2 images. This general agreement between CME speeds measured near the Sun, at the location of ACE, and at 1 AU supports our conjecture for a causal relation between this CME and the ensuing geomagnetic storm.

Observations from the LASCO C3 coronagraph show two CMEs on 9 August 2001 (one from the west limb studied here, and the other leaving the east limb at approximately 22:18 UT). The measured speed of the CME associated with the empty filament channel suggests that the interplanetary CME observed by ACE and WIND is caused by the CME from the west limb. However, the size of the ICME at the locations of ACE and WIND is significantly larger than one that can be inferred from a relatively short-duration geomagnetic storm. Also, the timing between, the ICME onset at the location of ACE and WIND and the onset of the geomagnetic storm, suggests that this ICME had largely missed the Earth. Thus, the geomagnetic storm observed on 13 August can be associated with one of the "threads" of the ICME observed at ACE and WIND. The complex magnetic topology of the ICME could be explained as the result of interactions between CME flux rope structure originating from the magnetic field over the filament channel and the magnetic field of the coronal hole.

2.2. Case 2: Eruption from a Filament Channel on 3 March 2011

A CME without a clearly identifiable source region on the disk was observed on 3 March 2011 by the two STEREO and the SOHO spacecraft. The separation angle between STEREO A and B was about 182° allowing observation of the CME from its two opposite sides while the front view was deduced from SOHO/LASCO data (see Figure 8). Similar to case 1, there initially was no clearly identifiable source region for this CME. By applying a geometric triangulation method to STEREO A and B data, we have estimated the heliographic coordinates of the source region as $\approx$ S35° $\pm$ 10°, W15° $\pm$ 10°. Figure 9 shows the derived area against SDO/AIA 193 Å, SDO/HMI magnetogram and Hα BBSO images. Additional inspection of this data around the triangulated heliographic coordinates places the source of the CME on a filament channel without a definite filament in the southern hemisphere. Figure 10 shows three consecutive images during the early and late stages of eruption. White arrows in Figure 10 indicate the two-ribbon flare-like emission developing in the corona on both sides of filament channel at the time of the overlying eruption. The filament channel is clearly identifiable in Hα images (Figure 11). On Figure 11 (upper panel), the filament channel can be seen as a slightly darker curved "path" that begins south-west

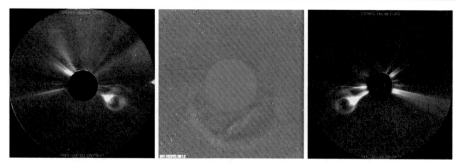

Figure 8 Left and right panels: the CME observed by the COR2 instrument onboard STEREO-B and STEREO-A, respectively, on 03 March 2011 at 09:09 UT. The simultaneous observations of a CME on both STEREO A and B in these orientations can only be due to an event with its origin near the middle of the solar disk. Middle panel: Half-halo CME observed by SOHO/LASCO C2 03 March 2011 at 09:12 UT.

(below and to the right) of bright active region plage, and "arches" below the active region to the east. On Figure 11 (lower panel), the location of the filament channel is marked by dark fragments of filament material. The background of the channel appears to be slightly darker on average than the surrounding areas (Figure 11, top) due to the relative absence of plagettes or bright areas in the channel. In images of higher spatial resolution than these, one would typically be able to see some of the fibrils in the channel. Especially near the filament segments and along the boundary between the opposite polarities, fibrils align with the polarity reversal boundary or have a component aligned with the polarity reversal boundary.

As the characteristics of observed changes are gradual and minor in the chromosphere and corona, we suggest there is gradual loss of equilibrium prior to the onset of this CME. As one possibility, the new flux emergence in the vicinity of the filament channel may lead to the destabilization of filaments contained within them (*e.g.*, Bruzek, 1952). The *Bruzek relationship* of specific erupting quiescent filaments to the birth of new active regions has been verified in subsequent studies (Feynman and Martin, 1995; Wang and Sheeley, 1999; Feynman and Ruzmaikin, 2004; Jing *et al.*, 2004; Balasubramaniam *et al.*, 2011). The eruption on 3 March 2011 happened at the time when a new strong active region began to emerge northwest from the filament channel. Figure 12 shows PFSS magnetic field line extrapolations in the vicinity of the filament channel before and after the new active region's emergence.

Our application of the model allows us to deduce whether the new flux could or does interact with the coronal fields that straddle the filament channel. It is clear from Figure 12, as new magnetic flux emerges, some of its field lines connect to preexisting flux of another active region. In turn, this leads to establishing new magnetic connections between the mature active region and one of polarities of magnetic arcade overlying filament channel. Thus, new large flux emergence could result in a change to the overall magnetic configuration around the filament channel and a weakening or decrease of the overlying coronal arcade field by transferring connectivity to the newly emerged adjacent magnetic flux. Similar changes in magnetic connectivity have been recently observed by Balasubramaniam *et al.* (2011) prior to a filament eruption. We should emphasize, however, that we use field lines extrapolated from the PFSS model only as a graphical representation of possible change in the topology. To properly model this development, a more sophisticated model would be needed that takes into account possible non-potentiality of the magnetic field around filament channel. It is worth noticing, however, a recent paper by Liu, Zhang, and Su (2011)

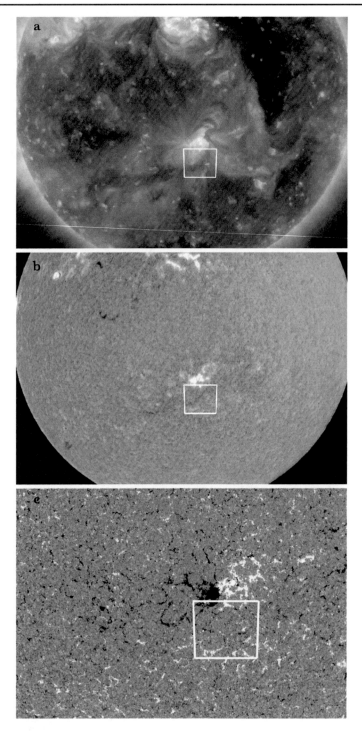

Figure 9 The source region of the CME in Figure 8 is identified with the filament channel and polarity boundary within the white boxes. From top to bottom (all on 2 March 2011): (a) SDO/AIA 193 Å at 18:00 UT, (b) BBSO Hα at 17:56 UT, and (c) SDO/HMI at 17:53 UT.

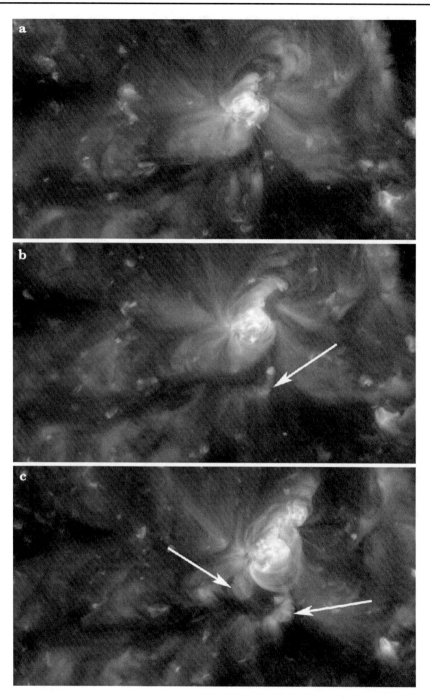

Figure 10 Brightenings reminiscent of a weak two-ribbon flare were observed after the slow CME eruption by SDO/AIA 193 Å: (a) 14:00 UT, 2 March 2011, (b) 21:15 UT, 2 March 2011, and (c) 09:15 UT, 3 March 2011. Unlike post-flare loops, these coronal structures on each side of the channel (shown by the white arrows) do not appear to straddle the empty filament channel although they develop together with a small transient coronal hole southward from channel.

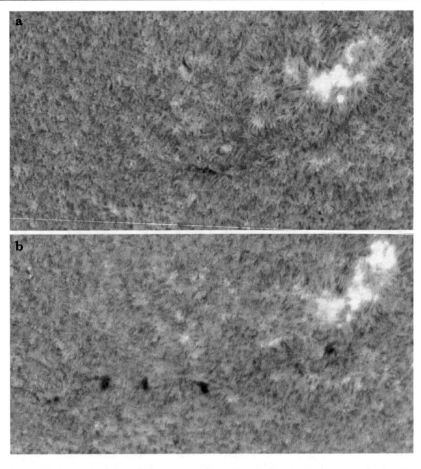

Figure 11 Chromospheric pattern of plagettes and fibrils along the filament channel (a) before and (b) after the flare-like brightenings and CME that both straddle this filament channel. Small fragments of the filament plasma are seen here inside this relatively empty channel observed in Hα (BBSO). Panel (a) shows data from 2 March 2011 at 17:56 UT, and (b) is for 3 March 2011 at 17:56 UT.

who compared the configuration of coronal magnetic fields from PFSS and non-linear force-free (NLFFF) models. They found that NLFFF and PFSS extrapolations agree reasonably well at the heights > 2000 km above the photosphere. Wang and Sheeley (1999) used the PFSS model to investigate the effects of a newly emerging flux on coronal arcades overlying filaments. They concluded that emergence of a new magnetic flux in the vicinity of filaments will weaken the coronal arcade above filaments and may lead to their eruption. The situation shown on Figure 12 appears to be in agreement with conclusions of Wang and Sheeley (1999).

Unlike the eruption of 9 August 2001 (our case 1), this CME did not produce an ICME at the ACE spacecraft location. No geomagnetic storm associated with this CME was observed either. Given the observed trajectory, we suggest that the CME passed the Earth orbit below the ecliptic plane.

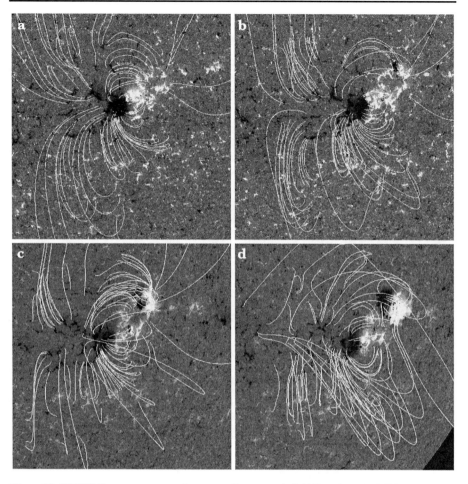

Figure 12 SDO/HMI magnetograms and corresponding magnetic field lines from the PFSS model before (panel (a), 2 March 2011) and during the emergence of the new active region (panels (b) 3 March 2011, (c) 4 March 2011, and (d) 5 March 2011. Time in all panels corresponds to 06:10 UT). While the PFSS model is not expected to exactly represent conditions of solar magnetic fields, it provides evidence that at least the depicted amount of change most likely occurred in association with the emerging active region and that these changes would be consistent with altering the magnetic configuration sufficiently to result in the flare in Figure 10 and the associated CME in Figure 8.

3. Discussion

Two examples described in this article indicate that CMEs can erupt from relatively empty filament channels, and that such CMEs can produce both geoeffective and non-geoeffective CMEs. In the absence of a clearly identifiable source region on the disk, successful detection of the origin of such "stealth" CMEs requires a wide and complex examination of the filament channels as candidate sources, together with the monitoring of new flux emergence in the vicinity of filament channels.

CMEs from magnetic systems encompassing filament channels without conspicuous filaments, similar to the ones described in this article, may explain CMEs that appear as having no obvious source regions on the solar disk. Close examination of EIT 304 Å images taken around the time of the event reported in Robbrecht, Patsourakos, and Vourlidas (2009) shows a compact brightening similar to footpoint brightenings characteristic of the phase of maximum acceleration of erupting filaments (Wang, Muglach, and Kliem, 2009). Also, after the CME, we see the formation of two elongated fragments of the chromospheric filament material inside the channel. These fragments are indicative of the location of the chromospheric filament channel, which apparently did not contain dense material prior to the CME. In our opinion, these post-eruption features suggest that the CME described in Robbrecht, Patsourakos, and Vourlidas (2009) could also have originated from a magnetic system with a filament channel without a conspicuous filament. For additional examples of CMEs without obvious solar surface source regions, we refer the reader to the events described in Bhatnagar (1996), McAllister *et al.* (1996), and Shakhovskaya, Abramenko, and Yurchyshyn (2002). Another more recent example is the CME eruption on 23 May 2010 at about 16:00 UT observed by *Solar Dynamics Observatory* (SDO). Preliminary analysis of data suggests that this CME may have also originated from a filament channel without an obvious filament.

Acknowledgements   The authors thank M. Velli for reading and discussions on the manuscript. A.P.'s work had benefited from partial support provided by NASA's NNH09AL04I inter-agency transfer. O.P. and S.M. were supported by NSF grants 0837915 and 0852249. National Solar Observatory (NSO) is operated by the Association of Universities for Research in Astronomy, AURA Inc. under cooperative agreement with the National Science Foundation.

References

Balasubramaniam, K.S., Pevtsov, A.A., Cliver, E.W., Martin, S.F., Panasenco, O.: 2011, *Astrophys. J.* in press.

Bhatnagar, A.: 1996, *Astrophys. Space Sci.* **243**, 105.

Bruzek, A.: 1952, *Z. Astrophys.* **31**, 99.

Cremades, H., Bothmer, V.: 2004, *Astron. Astrophys.* **422**, 307.

Feynman, J., Martin, S.: 1995, *J. Geophys. Res.* **100**(43), 3355.

Feynman, J., Ruzmaikin, A.: 2004, *Solar Phys.* **219**, 301.

Foukal, P.: 1971, *Solar Phys.* **19**, 59.

Jing, J., Yurchyshyn, V.B., Yang, G., Xu, Y., Wang, H.: 2004, *Astrophys. J.* **614**, 1054.

Heinzel, P., Schmieder, B., Fárník, F., Schwartz, P., Labrosse, N., Kotrč, P., Anzer, U., Molodij, G., Berlicki, A., DeLuca, E.E., Golub, L., Watanabe, T., Berger, T.: 2008, *Astrophys. J.* **686**, 1383.

Gaizauskas, V.: 1998, In: Webb, D., Rust, D., Schmieder, B. (eds.) *IAU Colloq. 167: New Perspectives on Solar Prominences*, ASP Conf. Series **150**, Astron. Soc. Pac., San Francisco, 257.

Gaizauskas, V., Mackay, D.H., Harvey, K.L.: 2001, *Astrophys. J.* **558**, 888.

Gopalswamy, N., Shimojo, M., Lu, W., Yashiro, S., Shibasaki, K., Howard, R.A.: 2003, *Astrophys. J.* **586**, 562.

Gopalswamy, N., Mäkelä, P., Xie, H., Akiyama, S., Yashiro, S.: 2009, *J. Geophys. Res.-Atmos.* **114**, A22.

Lin, Y.: 2004, Ph.D. Thesis, Institute of Theoretical Astrophysics, University of Oslo.

Litvinenko, Y.E.: 1999, *Astrophys. J.* **515**, 435.

Litvinenko, Y.E., Martin, S.F.: 1999, *Solar Phys.* **190**, 45.

Liu, S., Zhang, H.Q., Su, J.T.: 2011, *Solar Phys.* **270**, 89.

Ma, S., Attrill, G.D.R., Golub, L., Lin, J.: 2010, *Astrophys. J.* **722**, 289.

Mackay, D.H., Gaizauskas, V., Yeates, A.R.: 2008, *Solar Phys.* **248**, 51.

Martin, S.F.: 1990, In: Rudzjak, V., Tandberg-Hanssen, E. (eds.) *IAU Colloq. 117: Dynamics of Quiescent Prominences*, Springer, Berlin, 1.

Martin, S.F.: 1998, *Solar Phys.* **182**, 107.

Martin, S.F., Panasenco, O.: 2010, *Mem. Soc. Astron. Ital.* **81**, 662.

Martin, S.F., Livi, S.H.B., Wang, J.: 1985, *Aust. J. Phys.* **38**, 929.

Martin, S.F., Lin, Y., Engvold, O.: 2008, *Solar Phys.* **250**, 31.

McAllister, A.H., Dryer, M., McIntosh, P., Singer, H., Weiss, L.: 1996, *J. Geophys. Res.* **101**, 13497.

Panasenco, O.: 2010, *Mem. Soc. Astron. Ital.* **81**, 673.

Panasenco, O., Pevtsov, A.: 2010, In: Cranmer, S.R., Hoeksema, J.T., Kohl, J.L. (eds.) *SOHO-23: Understanding a Peculiar Solar Minimum*, ASP Conf. Series **428**, Astron. Soc. Pac., San Francisco, 123.

Panasenco, O., Martin, S., Joshi, A.D., Srivastava, N.: 2011, *J. Atmos. Solar-Terr. Phys.* **73**, 1077.

Pevtsov, A.A., Neidig, D.: 2005, In: Sankarasubramaniam, K.S., Penn, M.J., Pevtsov, A.A. (eds.) *Large Scale Structures and their Role in Solar Activity, ASP Conf. Series* **346**, Astron. Soc. Pac., San Francisco, 219.

Robbrecht, E., Patsourakos, S., Vourlidas, A.: 2009, *Astrophys. J.* **701**, 283.

Shakhovskaya, A.N., Abramenko, V.I., Yurchyshyn, V.B.: 2002, *Solar Phys.* **207**, 369.

Smith, S.F.: 1968, In: Kiepenheuer, K.O. (ed.) *Structure and Development of Solar Active Regions, IAU Symp.* **35**, Reidel, Dordrecht, 267.

Subramanian, P., Dere, K.P.: 2001, *Astrophys. J.* **561**, 372.

Vásquez, A.M., Frazin, R.A., Kamalabadi, F.: 2009, *Solar Phys.* **256**, 73.

Wang, Y.-M., Sheeley, N.R.: 1999, *Astrophys. J.* **510**, L157.

Wang, Y.-M., Muglach, K., Kliem, B.: 2009, *Astrophys. J.* **699**, 133.

Wood, P., Martens, P.: 2003, *Solar Phys.* **218**, 123.

Zirker, J.B., Martin, S.F., Harvey, K., Gaizauskas, V.: 1997, *Solar Phys.* **175**, 27.

Printed by Publishers' Graphics LLC
MO20120522